"Use Your Knife to Save a Life!"

KNIFE COMBATIVES
BY W. HOCK HOCHHEIM

Hardcover ISBN: 978-1-932113-40-2
Paperback ISBN: 978-1-932113-22-8
Digital ISBN: 978-1-932113-26-6

Copyright: First Edition 2009, Second Edition 2021
Published by Lauric Enterprises, Inc.
www.ForceNecessary.com
McKinney, Texas

All rights reserved.

The author, publisher, or seller do not condone or support terrorism anywhere, in any form. They do not assume any responsibility for the use or misuse of information contained in this book. The purpose of this book is to historically and artistically preserve the information contained within these pages for posterity. The book provides information that describes various methods of self-defense that may be employed against illegal aggression. In some cases individuals may make the choice to use their knife to save their life or the lives of others when it is morally, legally and ethically appropriate to do so. Providing the information about how to do this in no way condones or suggests that using a knife is ever justified. That decision is left completely to the individual and situation. Anyone who uses the techniques bears the responsibility entirely for any and all legal consequences of their independent actions.

Other Titles by W. Hock Hochheim
The Great Escapes of Pancho Villa
Fightin' Words
Impact Weapon Combatives
My Gun is My Passport
Last of the Gunmen
Rio Grande Black Magic
American Medieval
Be Bad Now
Blood Rust
Training Mission Series

TABLE OF CONTENTS

Author Foreword	5
Introduction: Mission Objectives	9
Chapter 1: Modern Knife Combat: The History and the Future	11
Chapter 2: The Knife Combatives Preparation Checklist	15
Chapter 3: The Weapons Range Continuum	17
Chapter 4: Tenacity of Life and the Knife	21
Chapter 5: Selecting Your Knife, Course and Carry	25
Chapter 6: Explaining the Drills	27
Chapter 7: Knife Grips	29
Chapter 8: Knife Stress Quick Draws	37
Chapter 9: The Knife Combat Clock	45
Chapter 10: The Knife Combatives Ready Positions	51
Chapter 11: Knife Attack Points	53
Chapter 12: Footwork and Maneuvering	55
Chapter 13: The Saber Grip Attack and Defend	65
Chapter 14: The Reverse Grip Attack and Defend	95
Chapter 15: Support! Hand Strikes, Grabs, Blocks and Kicks	115
Chapter 16: Quicker Kill, High Yield Targets and Attacks	125
Chapter 17: The Knife Passing Introduction	129
Chapter 18: The Stop 2 Ambush, Dodge, Evasion Exercise	131
Chapter 19: The Statue Drill Exercises	139
Chapter 20: Counters to Common Blocks Exercise	143
Chapter 21: The Knife Outside Invasion Series	145
Chapter 22: The Block, Pass and Pin Exercise	149
Chapter 23: The Windmill Exercises	153
Chapter 24: The Knife Horizontal Blast	157
Chapter 25: Knife Dueling	161
Chapter 26: Knife Grappling/Take Downs	177
Chapter 27: Countering the Weapon Threat Presentation	183
Chapter 28: Less-Than-Lethal Knife Combat Tactics	185
Chapter 29: Basics in Stand-up, Knife Combat Scenarios	187
Chapter 30: The Four, Knife Grappling, Combat Scenario Modules	229
Chapter 31: Addendum to the Knife	287

Author's Foreword

Knife fighting is not an art. Or a hobby. Or an interest. It isn't fun. Nor is a bullet to the brain or a broken leg with an exposed, splintered bone though the flesh. If you think it is, then I think you need your head examined or a real dose of reality experience.

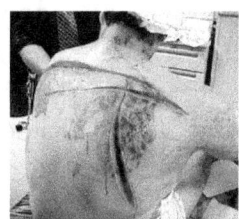

This is a book about the very essence of knife/counter-knife combatives. It's sources run the gamut from forensic science, crime, war, military, law enforcement and martial arts. It has been my lifelong pursuit to bridge the gap between these sources, as each group knows things about fighting the other doesn't. I am but a mere, dedicated vessel for all these points of reference. I myself have no great fondness for knives, sticks or guns. Instead, I am obsessed with tactics first and foremost. The truth in combatives lies behind martial systems into the generic, non-denominational strategy and tactics based on war and crime. Pull the curtains aside. See the back wall.

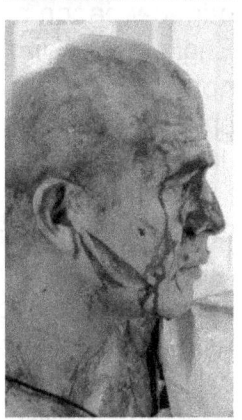

This book is meant primarily as a military resource book, with a heavy militant flavor and feel. You will read terms like "defender," "soldier" and "enemy" to designate the trainer and trainee positions and those terms. You will find numerous, true military knife combat stories and various military themes. This militancy to me is the core truth from which the soldier, law enforcement and the aware citizenry can find edged-weapon, ideas, methods and training ideology.

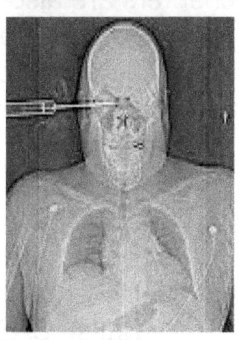

Under the scope and spectrum of this military filter, the most, effective and brutal methods can be explored without distracting concerns for humanity or political correctness. Life, death and tactics. It just is what it is. And from this violent base is the true vile nature of knife combat. Brutal. Ugly. Vicious. Disgusting. Wet.

It is my personal wish for you that you may never have to lift a knife for self-defense or defend any righteous cause: that you live a prosperous life without serious confrontation or violence: that you read this book only to prepare for a worst-case scenario. I demand that you use this information for the common good of oneself and humanity.

So, it is with some regret, some hesitation and trepidation that I present to you this very dark book. I do this because I know the nature of time and humanity. I know this is information that needs to be collected somehow and passed on by someone, somewhere so good people may survive the ebb and flow of evolution, and for the dark days that surely come and then surely pass again. Again and again.

We gather in such practice somewhat reluctantly. We pursue with an uncertain sadness, with purpose. As a surgeon uses a sharp scalpel to cut flesh and preserve life, I ask you to do the same with this knowledge. Use your knife to save your life, your family and friends, your proper government.

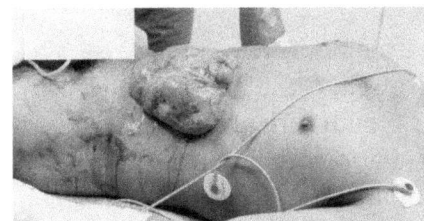 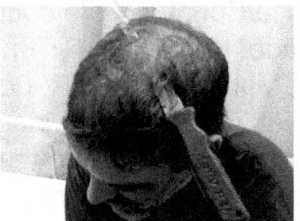 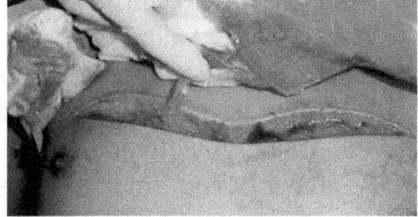

Since the 1980s, I've been collecting knife-attack newspaper clippings, articles and dog-earing pages of military biography and history books on the subject as I was interested in learning from actual experiences. To write this encyclopedia edition, in 1999 and 2000, I visited several military museums and libraries, and before and since, I have purchased some 500-plus military history books and biographies.

Between the libraries and purchases, I scanned biographies and histories to locate edged-weapon mentions or actual fights. Also please note that I have been limited to English-speaking and rare translated sources. I have always known that these non-English nations and their militaries have leaders in these fields, as we English-speaking peoples have ours. But, I have not been able to crack deep into the Chinese, Japanese, Indian, Spanish and other language histories, experiences and lessons learned. I mean no disrespect for this failing. But, I do know enough to report that the incidents and situations are universal in combat.

I would like to record here for the record, my scanning process which enabled me to speed-read tons of books. I quickly learned that the common military biography is set up as follows. The opening chapter is a teaser to capture your attention. The next chapters are the obligatory, "I was born down on a farm in..." I would scan over this. These few chapters are also about the enlistment and just how tough the basic and advanced training was. I scanned over, looking for knife training, usually finding none. Then to the front! The action begins. This is where I would slow-scan. Then to the battles that must be read very closely. This is the process of how I scanned all these books searching for knife training and knife combat incidents.

These military resources include:
United States Army manuals (old and newer).
United States Marines manuals (old and newer).
British SAS manuals (old and newer).
British Army manuals (old and new).
Finland Military Army manual.
Royal Thailand Army manual.
German Army manual (old and newer).
Turkish Armed Forces manual.
Australian Defense Force Special Operations manual.
Armed Forces of the Philippines knife manual.
Russian Army Ground Force manual (Soviet old and newer).
Manuals and special editions of various military units.
Military books from the 1940s through the present about military combatives.
Hundreds of military biographies from the 1940s through the present.

My Personal "Hands-On" Military Knife Training:
US Army Basic Training Knife course (Vietnam era).
US Army Military Police Knife/Counter-Knife course.
US Army Military "Police Judo" Knife/Counter-Knife training.
Regular training from ROK Marine Captain R. Lee (ret.) in South Korea.
Knife training from retired vet, Captain Ben Mangels South African Military.
Training with various martial artists who have taught various militaries.
Developing training courses for various military groups.

Multiple black belts - Hawaiian, Japanese and Filipino martial arts.
Multiple police-related knife courses.
Some 900 arrests, some involving edged-weapons.
Numerous police street survival schools.
Over 1,000 or more assaults, aggravated assault and murder investigations.
Over 20 assault and violent death schools and symposiums conducted by forensics experts.
40-plus years of study in and about edged weapons.
Interactions with the military as an instructor/paid consultant for all four branches.
of the US, German, British and Australian militaries.

And some last observations before we begin. I hope you find this book educational. It is indeed a college course on the subject, if not a masters degree, on the generic essence of knife combatives with hard-core military objectives at its base. For those seeking advanced skill and flow drills, be sure to study those sections. I firmly believe practitioners should train at the very least one skill drill per tactical step, to build performance depth. Most militaries do not have the time for this specific knife training. If they have the time? They do. Practitioners should not be overwhelmed or consumed by these "flow" drills, or mistake the drills for combat scenario training.

You cannot become a competent knife or otherwise "fighter" without a solid base in unarmed combatives. This is a book about knife combatives. This entire manual is based on you, the reader, the practitioner holding a knife in your hand. You will be fighting with your knife versus knife, your knife versus stick, your knife versus empty hands and your knife versus certain gun threats. It is possible that you may fight with your hands before you draw your knife, or you may loose your knife during the fight, and may have to fight unarmed. Or, pick up an impact weapon, or even a pistol. The foundation for knife "fighting" is first, unarmed combatives.

> **Black Box Knife Combat Files**
> *"I had a pet monkey that I kept with me in my hootch. Whenever there was an attack, this monkey would quickly climb on me, wrap his arms and legs around my head and start pissing on my face! I could not pry this damn monkey off my head. So here I was, running to the old French bunker that I used for a command post, with this monkey on my head, claws dug into my skull, pissing like hell on my face! Somehow, I just never got used to that!"*
> – Gerry Schumacher
> *War Stories of the Green Berets* - Motorbooks

INTRODUCTION: MISSION OBJECTIVES

Are you familiar with the grand experiment? The one in which "the butcher, the baker, the candlestick maker" were shoved to the front to fight...into the laboratory of horror called war? The test called combat? The real ugly monkey clawing onto their face, brain and body? It's the same one clutching on the backs and faces of soldiers, civilians, martial artists and police officers faced with combat. The fear of fear, the terror of being cut, gutted, maimed, shot or killed clings tight. And then there are the back room doubts.

All the doubts... can you leap over your natural, peaceful barriers into quick and deadly edged weapon violence? Can you shove aside your religion and your civilized opinions? Can you conquer hesitation? Can you discard human empathy? Can you always be alert enough when it counts? Can you cut your response time, increase reaction time, can you succeed beyond your lack of confidence and realistic training? The monkey list goes on and on, and for many it feels more like a 300 pound gorilla than a small monkey.

This textbook is dedicated to presenting an overview on the psychological aspects and physical actions of real knife military combat as seen by the many who experienced action first hand in the modern militaries of the world. I have explored hundreds of oral recollections about knife fighting in approximately 1,000 historical, military texts, and recorded many of these knife fights here. I did not search for books by generals, or books about campaign policies. Instead, I sought personal, grunt stories, trench-level reality. I selected only a few stories among many in each category that aspire to enlighten the reader on several differing levels and subjects. I want to take you to a higher place, to make you grapple with terror, heroics, shock and rage. I want you to face the brutal realities and dark, wet nature of maiming and killing. My mission is multi-layered.

> *First,* to prove that knife fighting has and always will exist in war as well as in low-intensity conflicts, or *Other Than War* (OTW) conflicts and policing actions.

> *Secondly,* I wish to review military knife training methodologies and disclose that most military knife training is inadequate for the realities faced.

> *Third,* to remind all trainers that while knife combat has and will exist, it exists inside a small probability of a larger weapons continuum.

And fourth, I will dissect these real military knife fights, filter the action through modern, tactical drills, strategies and techniques, and finally establish the optimum, winning methods for mental and physical survival. The scenarios in this book have actually happened. The solutions and progressive training outlines organized in this book are practical, tactical problem solving.

Mission Objectives

1: Knife vs. knife combat, though probable, is a small probability in modern weapons continuum. We will establish when and where knife fights exist in a mixed weapon and firearms world.

2: Current training is misguided and inadequate when compared to reality.

3: Review actual events, and construct practical, tactical solutions and doctrine training to confront the researched realities.

4: Turn a *raw-recruit to a combat vet* with progressive, modern training program that best prepares a soldier for the realities of knife combat.

5: Knife ground-fighting. A main purpose of this book is to bring recognition to the importance of knife ground fighting, a virtually ignored subject in martial studies.

The Knife Fighting Matrix

Knife versus unarmed attackers.
Knife versus impact weapon attackers.
Knife versus knife attackers.
Knife versus gun threats.

Standing.
Kneeling/sitting.
Grounded.
    - on back.
    - on top.
    - on right side.
    - on left side.

Chapter 1: Modern Knife Combat: The History and the Future

"Could you stick a knife in a man ... and twist it?"

"Could you stick a knife in a man ... and twist it?" was an early litmus test question of British Commando and U.S. Ranger recruits. "The anxious candidate would of course answer yes, knowing full well that the elite cadre needed such killing in their missions. "... and within a year," reports British Commando historian James Ladd, "he had done so."

But what is this *Military Knife Combat?* Where and how, in this age of 20th Century projectile firearm, weaponry, does it really occur that often?

At first, a naive researcher would assume that there is more than a considerable amount of knife fighting in war. Then, upon further examination, he deduces that soldiers are shooters first and foremost and there is very little, if any knife combat.

Within certain focused scopes of study, there are no knife reports and the knife is reduced to utilitarian uses in the military theater, like cooking, camping, shaving, mechanic work, and a host of other mundane tasks. Mirrors have been wrapped to its end and poked into the unsafe air to peer around corners and over trenches.

Knives have been traded for other services and products. On the more violent side, the knife has been used for torture and interrogation of prisoners, or buried tip-up, pungistick style and in other manners to arm various man traps.

However, a more thorough and diverse research into the oral histories and accounts of modern-day warriors around the globe reveals that military knife versus knife combat is a definite reality.

In some instances entire units fought each other knife vs. knife, and not always because of stealth or lack of ammo. Although that has happened, but because close-quarter fire fights would accidentally kill their own intermingled comrades.

In the trenches, alleys, foxholes, and ruins, houses and bunkers of war, and in the drama of its espionage, knife combat has always existed. Whenever and wherever a soldier needed silence, or ran out of bullets, or couldn't reach his rifle or pistol (the majority are not issued pistols), or couldn't shoot due to proximity of comrades and explosives, the knife has left its scar or corpse.

Documentary news footage records men like Panama's Noriega and Iraq's Saddam Hussein admiring swords and knives and at other times brandishing them in the air before political rallies. Hundreds, if not at times thousands of frenzied soldiers waved their issued swords and knives back! Hard to imagine one of our USA Pentagon generals or allied politicians leading such a pep rally.

In largely agrarian societies around the globe, as in the Archipelago, South America, Africa and Asia – all hot spots for terrorism, riots and revolution, a familiarity with edged weapons like bolos/machetes, swords and knives has made their presence natural in standard military issue, training and combat.

Without a doubt, a cultural edged-weapon presence can influence the type and behavior of combat and modern ground troops do not always fight bullet versus bullet against the well fed and well equipped. Vets recall that WW II basic trainees were instructed –

"If you're gonna' fight Germans, you may never use a knife, but I guarantee if you're gonna' fight the Japs, you'll be using a knife in combat."

The prevalent sword mentality of Japan had Imperialist Troops charging Marines and rattling sabers under gunfire in the South Pacific Theater of WW II.

Despite the possibilities of edged weapon combat throughout our world, despite the proliferation of its militant image, military administrations muster little effort prioritizing military knife training programs.

This research course will review historical tactics, advice and programs, review actual knife encounters, move into an evolutionary, modern science of physical and mental performance, and then meld it all together to produce innovation and enlightenment for all combatives disciplines.

Black Box Knife Combat Files

The prevalent use of the machete, swords both long and short, and the knife, is always higher amongst the ill-equipped armies and rebels. Any kind of ambush, rushing, close quarter, edged-weapon attacks may often prohibit fire fights, as evidenced in this 1980's television documentary –

"We had nothing. Just machetes and some knives ... some sticks. Shovels. Maybe some had some old guns of some kind. The police (military) were coming in a lazy row. More than ten? Fourteen? As they passed, we jumped out and cut them down ... They did not shoot once because they were not ready to shoot amongst them and us. I stabbed one in the chest, and he falls, and I stabbed him again."
– *"South America-Our Neighbors Rebel"* Documentary-Television

Combat Notes
– Edged weapon cultures have access and familiarity with long and short knifes.

– An ambush rush into the enemy limits their willingness to fire in and near their comrades.

– A lack of proper alertness/attention by the marching troops.

– The use of improvised weapons like shovels.

Black Box Knife Combat Files

If you research the case histories of Medal of Honor winners since WW II, you would learn that about one of every 15 or so are stories of brave men who fought their enemies in close-quarter combat with their knives. Inside these heroic records you'll find the winners often surrounded by others in edged-weapon combat, such as WW II Medal winner U.S. Marine Capt. Louis Wilson –

"Seven times the Japanese charged (with short swords, knives and fixed bayonets.) Seven times Company F, under Wilson's indomitable leadership, fought them off. Several times...brave men wrestled the Japanese, using bayonets and knives to kill them."
— Edward F. Murphy, *Heroes of WW II*, Ballantine

Black Box Knife Combat Files

"He would use his knife."

"If there was a sudden rush from a large enough group, Howe felt, they could be overrun. He began preparing a check list for himself, the steps he would take in a final fight. He was going to take as many of them with him as humanly possible. He still had six or seven magazines left for his CAR-IS, along with his .45 and some shotgun ammo. He would shoot his rifle until it ran out of ammo, then the shotgun, then his pistol, and finally he would use his knife."
— Mark Bowden, *Black Hawk Down*, Atlantic Monthly Press

Paul Howe
www.combatshootingandtactics.com

(One of the best firearms instructors to be found anywhere. Located in east Texas.)

CHAPTER 2: THE KNIFE COMBATIVES PREPARATION CHECKLIST

A combat intelligence analyzer must ask and examine the six main questions to properly prepare a military assignment, a police mission, or a shopping trip, from a large-scale invasion to a small, one-on-one encounter. Those questions are: who, what, when, where, how and why. The answers range from big concepts on down to the smallest detail. This should be an unforgettable formula for every aspect of planning.

I have written so much, so many places about the who, what, where, when, how and why of the knife, I will only quickly overview the questions here. A very brief sample follows.

I like to advise people in seminars that they must answer these questions in the "micro" and the "macro," as in the small and the big. It takes about six passes to get a handle on the list, because in early stages you realize you have to go back and add and investigate the questions again.

I ask you to read my *Fightin' Words* book and the *Training Mission* Series books to cover the big picture of these Q and As.

Combatives Question One: Samples of Who?
Who are you to carry a knife? Who is he or she?
Who is the enemy? Who are you fighting?
Who are you exactly? Are you a prepared soldier, police officer or citizen?
Who will teach you how to use a knife?

Combatives Question Two: Samples of What?
What is your lifestyle?
What kind of knife?
What is going on? A covert mission? A patrol? A raid? An invasion?
Citizens in their everyday life? Knife-related sample questions:
What kind of knife will you have?
What will the enemy have? What will you do with your knife?
What will the enemy do?
What are the rules of this engagement?
What knife course will you take?

Combatives Question Three: Samples of When?
When will this knife incident happen? Day? Night? What season? What weather?
When and at what point in the tempo of the battle will this happen?
When will you use your knife?
When will your enemy use his?

Combatives Question Four: Samples of Where?
Where will this happen? A beachhead? A jungle? Water? A desert? An alleyway? A trench? A street?
Where will you carry your knife? Where will the enemy carry?
Where will you use your knife? Where will the enemy?

Combatives Question Five: Samples of How?
How will this happen. How will you attack? Defend? Surprise?
How will you draw, hold and use your knife? How will the enemy?

Combatives Question Six: Samples of Why?
Why is this happening?
Why will you draw your knife and use it? Why will your enemy draw?
Why are you "down to the knife?"
Why are you still there?

If planners take the time to answer these questions from the biggest concept on down to the smallest detail they will have successfully planned for a proper military, police and citizen action. Adding variables such as the knife into this examination will fine tune the training and improve performance.

Black Box Knife Combat Files

"Modern warfare consists of the massing of gigantic, agencies for the slaughter of men by machinery. Killing is reduced to a business like the Chicago stock-yards."

– Sir Winston Churchill quoted in Ron Clark's *Rise of the Bofffins* ("boffins" were scientists working in the miliary)

CHAPTER 3: THE WEAPONS RANGE CONTINUUM

The military knife by its very nature and size is a short range, close quarters weapon of silence and surprise, threat, control, shock and awe, injury and death. Measured in feet not yards. It exists near the end of an extensive, military weapons continuum and therefore sees little actual, practical use in combat in comparison. Since physical distance and range is an issue, here is a loosely contrived military weapons continuum by range and distance. keeping in mind that some might be interchangeable, based on situations.

Range 1:   Long-range missiles.

Range 2:   Short-range missiles.

Range 3:   Shoulder-fire missiles.

Range 4:   Long-range rifles (as in sniper).

Range 5:   Lessor-range rifles.

Range 6:   Grenades.

Range 7:   Handguns.

Range 8:   Bayoneted rifles.

Range 9:   Impact weapons.

Range 10: Knives.

Range 11: Other – Improvised weapons.

"*Down to the Knife,*" is an old military phrase. How often does a soldier actually engage in a knife versus knife fight? Despite the weapons continuum, It has, can and will happen and therefore knife versus knife, knife versus hand, knife versus stick, knife versus gun threats, and knife versus improvised weapons training must be properly conducted in relation to its situational possibilities.

Black Box Knife Combat Files

"This is the End."

NAJAF, Iraq – One of his friends was dead, 12 others lay wounded and the four soldiers still left standing were surrounded and out of ammunition. So Salvadoran Cpl. Samuel Toloza said a prayer, whipped out his knife and charged the Iraqi gunmen. In one of the only known instances of hand-to-hand combat in the Iraq conflict, Toloza stabbed several attackers who were swarming around a comrade. The stunned assailants backed away momentarily, just as a relief column came to their rescue.

Corporal Sam Toloza and the knife

"We never considered surrender. I was trained to fight until the end," said the 25-year-old Toloza, one of 380 El Salvador soldiers whose heroism is being cited just as criticism is leveled against other members of the multinational force in Iraq.

"We didn't come here to fire a single shot. Our rifles were just part of our equipment and uniforms. But we were prepared to repel an attack," says Col. Hugo Omar Orellana Calidonio, a 27-year army veteran who commands the Cuscatlan Battalion. *"We came here to help and we were helping. Our relationship with the people was excellent. They were happy with what we were doing,"* Calidonio says.

Then came April 4, when armed followers of radical Shiite cleric Muqtada al-Sadr seized virtual control of the city and staged attacks on two camps, adjacent bases on the fringes of Najaf occupied by the Salvadoran and Spanish units.

When Toloza and 16 other soldiers arrived that morning at a low-walled compound of the Iraqi Civil Defense Corps, about 1.2 miles from their camp, they found its 350 occupants had melted away and themselves trapped by al-Sadr's al-Mahdi militia. Lt. Col. Francisco Flores, the battalion's operations officer, said the surrounded soldiers held their fire for nearly half an hour, fearful of inflicting civilian casualties, even as 10 of their number were wounded by rocket-propelled grenades and bullets from assault rifle bullets.

After several hours of combat, the besieged unit ran out of ammunition, having come with only 300 rounds for each of their M-16 rifles. Pvt. Natividad Mendez, Toloza's friend for three years, lay dead, riddled by two bullets probably fired by a sniper. Two more were wounded as the close-quarters fighting intensified.

"I thought, `*This is the end.*' But at the same time I asked the Lord to protect and save me," Toloza recalled. They were out of ammunition. The wounded were placed on a truck while Toloza and the three other soldiers moved on the ground, trying to make their way back to the base. They were soon confronted with al-Sadr's fighters, about 10 of whom tried to seize one of the soldiers. *"My immediate reaction was that I had to defend my friend, and the only thing I had in my hands was a knife,"* Toloza said. He charged the Iraqi gunman with his knife! (*contin...*)

"As reinforcements arrived to save Toloza's unit, the two camps were under attack, with the El Salvadorans and a small U.S. contingent of soldiers and civilian security personnel trying to protect the perimeter and retake an adjoining seven-story hospital captured by the insurgents. U.S. troops have now replaced the Spanish, Salvadoran officers, many of whom were trained at military schools in the USA."
– Yahoo News

Combat Notes:
- This was knife versus guns.
- One versus multiples.
- Help arrived immediately.
- Folding knife not fixed-blade knife.
- Toloza managed somewhat of a surprise knife attack on busy troops.

Black Box Knife Combat Files

"Down to the utility knife."

"Close-quarter battle commences with the rifle. If this fails, the traditional backup is the pistol. Out of ammunition, or with no other alternative, the soldier falls back on edged weapons, starting with the bayonet. A step down from his bayonet is the entrenching tool as a cleaver. Down again is the utility knife."
– Jim Shortt, British Paratrooper and Combat Instructor

CHAPTER 4: TENACITY OF LIFE AND THE KNIFE

Through the years as a patrolman and criminal investigator in the US Army and on a Texas police force, I have worked numerous edged-weapon, assault cases that resulted in the full spectrum from threats, minor wounds, significant wounds, major wounds and lethal results. Hundreds and hundreds of them since the 1970s. In these last years since my "retirement," I have acted as a paid consultant, expert in knife crime for both prosecution and defense attorneys from California to North Carolina and states in between. And, I have never stopped my research in criminal and military knife attacks, as evidenced by the dozens of real-world, knife fights I've chronicled here in this book.

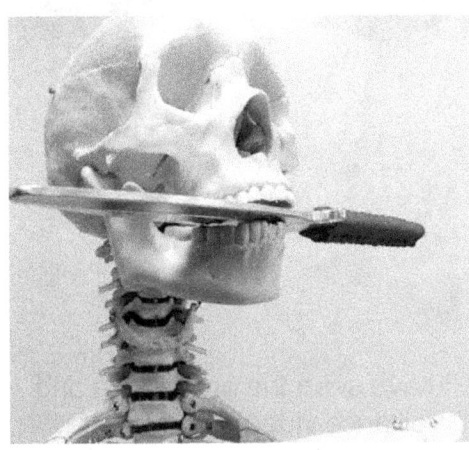

Probably the most educational experience I've had came from numerous investigation schools, such as annual *Assault and Violent Death* symposiums. These gatherings were taught by chief medical examiners from various major cities in the US and with lectures, slide shows and videos about the numerous autopsies and cases they'd worked. Many of these cases involved edged weapons, usually kitchen knives, as throughout the civilized world, common, single-edge, kitchen knives are used almost 99 percent of the time in crimes. All of these cases describe the various situations of criminal assault and the "who, what, when, where, how and why" these knives get pulled and used. Knife fights, all fights actually, are highly, highly situational. Universal themes, but very situational.

One point of universality, and points of contention amongst the ignorant is what does a knife attack really do to someone? How will a person react to a stab or slash both mentally and physically? The ignorant will proclaim all knife stabs kill, or that all slashed people cut wide apart like slabs of supermarket, roast beef hanging on a string, or that as soon as people are first cut or stabbed they experience medical shock and immediately debilitate. Or, that when people are hit from a certain direction with a knife, they will next, automatically, autonomically, bend in a certain, predictable, mandatory positions, perfectly setting up the next attack. Sort of like a game of billiards. Suffice to say that none of the above attacks will create absolute results each and every time. You can hope. You can plan. Bet. Assume. But you cannot guarantee. That is why I use the terms "high yield" and "probability factors" when discussing the effects of some tactics.

Much of this misinformation is brought to you by "the experts," many of whom are high school dropouts with just enough pocket money to advertise their name (an easy feat these days in the age of cheap web pages). Actually the experts come in all shapes and sizes beyond dropouts. We see, rough-tough, ex-military types with no real knife education and experience with the knife. We hear about experts who grew up on the "wrong side" of town or "tracks," or martial arts black belts with a good imagination, or drug addicts. One of my favorite "qualifying " backgrounds was from a guy who worked a midnight shift in an inner city, factory, and saw a lot of knife crime on his lunch breaks! How about people who did a little time in a county jail yet pretend to be tough, ex-cons and prison violence experts? What about the martial experts who build mountains out of mole hills? That is, mountain ranges of drills and moves out of simple mole hills of stabbing? Speaking of mole hills, a knife system can also be painfully, dangerously, over-simplified. Minimalists can go way too minimal. And, some of the inexperienced remain pristine from reality and just collect all of the above opinions and moves and manufacture their own programs. Houses of cards built from other houses of cards. What if a stiff wind comes?

One stiff wind is – as with firearms – that complete amateurs do lethal damage with knives. Simply put, simple knives with simple moves do much damage, and technical problems with mountains of tactical doctrine are easily forgiven and therefore easily overlooked because the margin of doctrine error is forgiving and broad. You might say, many knife experts get lucky and retain their title as one, just because a killer used a slash or a stab in a saber or reverse grip.

What will a knife do to you? Will the touch of the tip or the edge shut you down into shock and a pattern of predicted, autonomic responses? Of course not. This brings us to the core subject here of human nature and the continuum of the tenacity of life. I have seen the predictable and the unpredictable. In the early 1980s, I was shown a surveillance video of a vendetta, prison cafeteria knife fight. An ambush. In preparation, the killer stole two butter knives from the kitchen. He sharpened them into points and edges. With the help of an accomplice he duct-taped the two knives into his two hands. Often such prison killers, remove their shirts and swab their upper bodies with kitchen lard, oil or soap to reduce the chance of being grabbed. This guy didn't bother with that.

The killer found his target and closed in. Other inmates in the cafeteria scattered. The target could have grabbed his food tray as a shield against the knives, but he didn't think of it. I watched the film intently to see this victim fight the double knife attacker. The whole fight lasted but a minutes or so. The victim did the best he could but, all he knew to do was try and reflexively block and pass the knives and throw a few punches and slaps at the attacker's head. One thing in his favor was that he dodged and ducked all over the place. The tables, with attached benches, were very long and heavy an seemed to be bolted to the floor. But, footwork was on his list of survival and he ran, leaped and dodged the knives.

The target eventually died before guards could respond. The state investigator reported to us that the target was stabbed and slashed - I want to say from my memory - 187 times! If that is not the exact number, then it is close as it has been near 25 years since I saw this video tape. He noted that for the first 80 or 90 stabs and slashes, the target fought and moved very quickly and quite well. But the following 80 or so, he was losing it. Losing his wind, his balance and probably his blood pressure. Interesting point – the first 80 or so wounds took about the first 60 percent or 70 percent of the fight time. The following 80 or so wounds took the final 30 percent of the time. All in direct relation to the victim's inability to sustain his defense. He fell back and was mauled by the attacker. Guards charged in with night sticks and beat the attacker.

The video tape ended. The investigator said that the attacker, already doing life for murder, was charged and convicted of yet another murder. The state, unlike Texas, had no death penalty.

One hundred-plus stabs and slashes! That was a lot of knife work. I have worked cases where victims were stabbed and slashed many times. The worst I think was about 40 times. Some actually lived, and then some died. But the moral of all those stories is how unpredictable the stab/slash death toll/ratio can be. And, how unpredictable the responses to being stabbed or slashed were. In the jail film, the victim was moving and this movement interrupted the common responses one would guess a stabbed or stabbing person would do.

My old friend and biker comrade Gary Tile had his throat slit wide open one night in the 1970s, in our old New Jersey haunts. Gary was waiting at a red light in a bad "hood," when somewhat of a planned riot unfolded on the street. Residents rushed out to cars waiting for the red light. Gary said that car windows were broken and people were hauled from their cars and beaten. He tried to maneuver his motorcycle back and away from the car in front of him, so as to bust down the street or even down the sidewalk to escape. This plan worked...for a bit. A man with a knife in his hand saw him and lunged the blade out as he passed by, ripping the right side of his throat wide open. Gary continued on, touched his throat and saw the blood. He gripped his neck and lowered his head down to the right. Since it was our old neighborhood, he knew right where the local hospital was. Letting go to shift gears, he raced to the emergency room. He staggered in, a bloody mess, and was treated immediately with many stitches. As they worked on him, he saw the other victims of the riot being stretchered in.

Still, even with his throat severely slashed open, covered in blood, he made it to the hospital and was saved. Perhaps an eighth or quarter of an inch either side, he might not have managed to make the drive?

And then, I have seen a single stab kill a person, with a very small pocket knife! Or, a single slash in the right place can kill a man. I've seen this in person, not just in case files or films. I have chronicled in earlier blogs the night I arrested a murderer seconds after he stabbed a man through the eye and into the brain with a small pocket knife. Or the case I was called to of a teenager stabbed right in the heart by a small pocket knife. Small, novelty knife. Yet quick death. I remember the scene. A frigid, cold night. The teen had on a thick jacket, not thick enough, laying there on the gas station parking lot. Deader than hell. His face contorted in an expression of surprise.

Anyway, all these experiences, my cases, the case reviews of thousands of others, and research into military action and crime, tell me that any knife stabs or slashes may or may not:

    a) produce predictable, body responses and
    b) produce quick death.

The tenacity of life is the unpredictable force that keeps a victim, an opponent, or you, fighting on. It proves that a knife attack doesn't carry the shocking feel of a training stun knife, or a stun gun when touched (this type of electric training is dubious at best). If you are stabbed or slashed, do not give up!

You may well survive. If you do die, we fully expect you to last long enough to kill the bastard that killed you. Don't let us down on this point. The last thing, your last duty is to kill your killer, if you're going down. Take him with you.

Black Box Knife Combat Files

"The Ghurkas became noted for being exceedingly proficient at slicing off an enemy's head with one swish of a kukri blade...as in the old Gurkha story of a one-to-one with a Japanese officer armed with a fine samurai sword and swearing that no stumpy kukri-knife could match his glistening katana-weapon as he slashed the Gurkha in the arm and then cut off his hand.

'Ah,' replied the Gurkha, 'I may be wounded, but you, sahib ... I suggest that you do not nod your head!'

"As we advanced through the foothills which were dense trees, not jungle, we had a series of brisk actions against the Japanese. One I well remember ended in a rather macabre way. It was an afternoon battle. D Company and my own B Company were involved. It was very close fighting and was the first time that I heard Gurkhas actually shouting 'Ayo Gurkhali', which is a very fearsome noise at close quarters and undoubtedly scared the Japanese. It was also the first time that I'd seen them using their kukris at close quarters. They use them very briskly and mostly went for the throat, often putting down their rifles and bayonets. The Japanese ran, but we suffered quite considerable casualties.

They were not heavily dug in, and the Japanese never ran when they were heavily dug in. They used to see it out to the end. In this case they had only dug small foxholes which were in the form of trenches suitable for one man at a time. We killed probably 20 Japanese.

And the next morning I was instructed to take two platoons back to this position to bury the dead. The macabre part of my story is that I instructed the Jemadar to see that the bodies were buried, and I went off for some other purpose for half an hour or so. Rigor mortis had set in on the bodies left behind by the Japanese and would not easily go into the foxhole trenches dug by the Japs. When I came back, I found that instead of digging further graves, the Gurkhas were in the process of cutting up bodies and pushing them in. Of course, I had to stop this. It in no way concerned them, however, and they thought that I was being persnickedy, I think.

Goose Green itself was a shambles. There were 500 disgruntled Argentinian prisoners locked up in sheep-shearing sheds who were very frightened that their new guards were Gurkhas, having read the magazine that featured an article headlined 'Los Barbaros Gurkhos' which described in sordid detail how Gurkhas took no prisoners. These dispirited men were flown out to San Carlos and the 'barbarous Gurkhas' could then get on with the job at hand.

The victorious men of 2 Para were exhausted after an epic battle and many of its soldiers suffered from trench foot, a condition common during the First World War although rarely seen in modem times. The Paras were to move forward, leaving the Gurkhas to clear up the combat battlefield, construct new defensive positions and dominate the whole of the southern half of East Falkland by using a series of short, sharp, heliborne patrols. A patrol from the Reece Platoon ran into an Argentinian group armed with antiaircraft missiles. The Gurkhas forced them to surrender by simply drawing a single kukri and brandishing it aggressively. Later that kukri was to be auctioned to raise money for charity and fetched £300!" – The Gurkhas, John Parker

CHAPTER 5: SELECTING YOUR KNIFE, COURSE AND CARRY

Here are some important factors and legal issues, ignorant people fail to think of when buying a knife, or attending a knife course.

* Big knife or small knife. How big is big? How big do you need your knife to be?
* Conducive to a saber grip? Reverse grip? Both? (I suggest both.)
* Folder vs. fixed blade. It's a personal preference but make it an educated one. Once one is opened, it essentially becomes a fixed blade. And, with inventions like the Emerson "Wave," these folders are easier to open than ever. The folder opening mechanisms to experiment with:
  - the mentioned "wave": or hook that catches the corner of you pocket when drawn.
  - gravity or sling.
  - switchblade/spring-loaded.
  - hole or disc opening?
  - will your folder "button-pop open" in your pocket? Worry about these "auto-button openers" opening inside your pockets.
  - the "stealth/quiet draw" of the folder. Can you quietly open your knife without a click of some sort.

* Worry about a violent name of your knife and/or knife course. They will work against you in court. There are several horribly named courses lurking out there. Some have dark and violent messaging, gear and logos. They think they are macho, but they are fools. They are legal, bear traps to prison.

* Can your clip be screwed into the four corners of a folder knife? Citizen, milliary or police different clothing and missions cause you to carry weapons on different carry sites on your body. This effects your draw and your favored grip.

* Handle issues:
  - an ergonomic-shaped hilt/guard and handle is an absolute must to prevent hand slippage onto the blade.
  - maximum rough texture. Texture defeats blood, sweat, water and oily substance grip problems. Beware the "shinney," smooth and flat handles. Texture! Texture!
  - flat vs. round handle – the old commando knife handle problem. Their original handles were round and they could not tell by feel, where the edges are.
  - can you punch with a folder closed and not hurt your hand? This is a non-lethal application of knife use, that is very wise doctrine for a knife course. Sometimes you do not have the time to open the knife. You pull it, and must take immediate action, such as punching. A closed folder should support a punch.
  - does the folder ends protrude at the top and bottom of your fist? Like a palm stick, for closed folder, hammer fist, impacts? (For the reasons just mentioned above.)

A Training Tip for Classes and Seminars

If you're attending a knife class or knife seminar, and you bring a fixed blade trainer with you, you should also bring a belt and a sheath for it. You need to have a realistic carry system with you, so you can practice a realistic draw. These knives don't just appear like magic in your hand. You can buy cheap, "universal" sheaths that allow for various shape knives if need be.

Can You Carry and Use a Knife?

In some jurisdictions, in some cities, counties, states and countries around the world? Yes. You must first check the jurisdictions where you live and where you travel. Check laws on "brandishing a weapons," as discussed in a prior chapter. Another common, legal issue directly related to knives is the legal length of the blade. The fighting use of a knife is based on your local self defense laws, often very tricky.

The Three Knife Carry Sites

For fixed knives and folding knives, know all about the three weapon carry sites for yourself and your enemy. They are:

1: *Primary carry sites:* Think easy weapon quick draws, like the belt line and armpits. Pockets are also very quick draws.

2: *Secondary carry sites:* Think back-up sites. It takes some "digging" to get them out.

3: *Tertiary carry sites:* Think lunge and reach "off-the body."

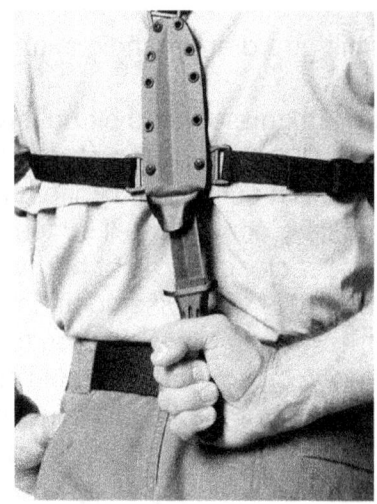

The very slim nature of knives, allow for some surprising carry sites.

A sheath in a pocket, attached to a string on a belt. Pull out on the handle. The sheath will come off in a "quick draw."

Chapter 6: Explaining the Drills

Foundational Athleticism

The soldier maintains an overall fitness level of endurance, fitness, speed, strength, explosive power, with a consistent program of:
- stretching.
- balance exercises.
- jogging long distances.
- wind sprinting.
- weight resistance or regular weight lifting.
- performing functional strength exercises that relate to fighting.

Solo Command and Mastery

The soldier exercises the prescribed movements solo, or alone and "in the air" to ensure that movements are kept efficient and there is overall power with balance and body synergy. The soldier then strikes training objects such as bags, shields, war posts, etc to experience realistic impact and generate maximum power. The soldier should visualize attacking an enemy when working alone. Establish a program of exercises that includes:
- 1: sets of ten movements each.
- 2: both right and left handed.
- 3: standing, kneeling, kneeling-over, seated, and on the ground.

Solo Command and Mastery. Hitting training objects. Working alone, here with impact.

Partner Drills: Feeding Targets

Next, a trainer will feed focus mitts, shields or sticks and the soldier strikes them. First standing and stationary. Then the trainer will move and feed, developing more target acquisition, footwork, power and balance.

Feeding the focus mitt is a skill. Start with the face of the mitt concealed. The trainer turns and flashes the face at the soldier. The shorter the flash time, the faster the soldier will become in hitting the mitt, be it a stab or slash. The same is true with holding a target stick. The trainer contracts the stick close to his body, then shoves it out at the four basic clock angles. The faster he produces and retracts the stick, the faster and harder the soldier strikes out. The mitt is best used to develop stabbing. The stick is best used to develop slashing.

Feeding and Hitting Drills.
    1: Stationary set.
    2: Moving set.

Establish a program of exercises that includes:
    1: sets of 10 or your choice.
    2: right and left handed.
    3: standing, kneeling (seated) and on the ground.

A trainer holds a stick to develop slashing and a focus mitt to develop stabbing skills for combinations.

Each basic tactic presented in this manual should be worked through the following exercises chart.

Solo Command and Mastery
    Exercise you strength.
    Exercise your endurance.
    Exercise your balance.
    Exercise your footwork & obstacle course work.
    Exercise your target acquisition.
    Work the tactic alone with no impact.
    Work the tactic alone with impact.
    Maximize body synergy.

A trainer holds a stick to develop slashing and a focus mitt to develop stabbing skills for combinations.

Partner Drills
    A trainer feeds stationary targets.
    A trainer feeds moving targets.
    A trainer feeds moving targets and various "flak." This means a trainer introduces some attacks and counter attacks to develop the trainee's skills.

    Skill developing drills.
    Combat scenarios.

At times shields can be used.

Note: *I strongy suggest you research Dr. Bill Lewinski's Force Science for deeper teaching and training methods, that cannot be covered in this theme book. www.ForceScience.org*

Right now, someone, somewhere is exercising and training hard to kill you.

CHAPTER 7: KNIFE GRIPS

First off and for the record, there are only two ways to hand-hold a common knife. One is the saber grip when the blade protrudes out of the top of the hand. The other is the reverse grip, when the blade protrudes out of the bottom of the hand.

A grouping of fixed blade Swedish military knives

What is the knife grip made of? What is the texture on the handle? What is the soldier's hand grip on the handle of the knife – the application of his fingers, palm, thumb and ball of the thumb upon the grip? Does the blade protrude from the top or the bottom of the hand?

1: What is the handle made of?
2: What is the texture of the handle?
3: How does the hand hold the knife handle?
4: Was the knife itself made by ignorant knife makers? Is the design maximized for combat?

Until a practitioner stabs a training object with force, he or she simply does not know of this common, combat risk of easy hand injury. Military knife combat involves blood, sweat and fluids of substances of the battlefield. Mud, water, slime, oil, whatever will quickly become an issue. These all inhibit the soldier's safest and effective grip on his knife.

Three things protect your knife-bearing hand in actual combat. The knife handle "hilt", the texture of the handle and/or the shape of the handle are mandatory design features of the combat knife.

First is the hilt/guard. The hilt prevents your hand from accidentally sliding up from the handle onto the blade when you stab the knife. In knife versus knife combat, the hilt offers the classic protection against the other knife striking the soldier's hand.

In some fixed and folding knives the very design and shape of the handle may prevent or reduce this dangerous hand slippage upon the blade.

Finally, the very texture of the handle itself plays an important part in retaining the knife and protecting your hand. Ridges and cross-cut indentations may offer enhanced grip. On the modern marketplace, various heavily textured tapes may be easily bought and applied to knife handles to improve grip.

The Wandering Thumb

For many years I have used the concept of the "wandering thumb" to describe the many positions atop or alongside a knife that the thumb may take to facilitate a saber or ice pick stab or slash. The diagram to the left originates from common meat butcher training, an excellent resource for knife combat training information, given the butcher's daily duties of working though bone, muscle and fat. The thumb and the ball of the thumb is vital in

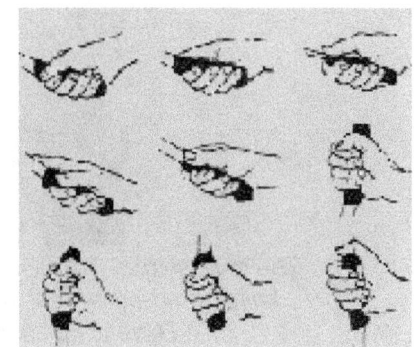

grip integrity. The thumb often "caps off" the knife handle pommel if the pommel is not itself designed to be a sharp weapon such as with the V-42 knife pommel's impact feature (as shown in the upcoming combat files box.).

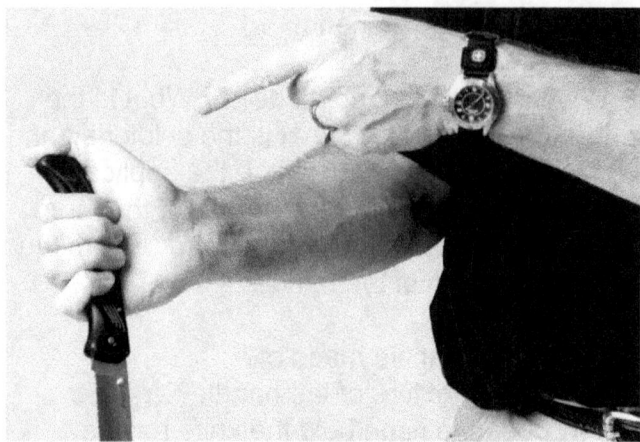

Using the thumb in this "cap-off" manner supports stabbing and as will be displayed in the upcoming slash section, the thumb really facilitates the reverse grip slash by fanning the knife forward into a target. This also helps extend the reach of a reverse grip stab.

It is not a good idea to have spikes or other designs that prohibit thumb support on the pommel. Flat is also a good striking surface.

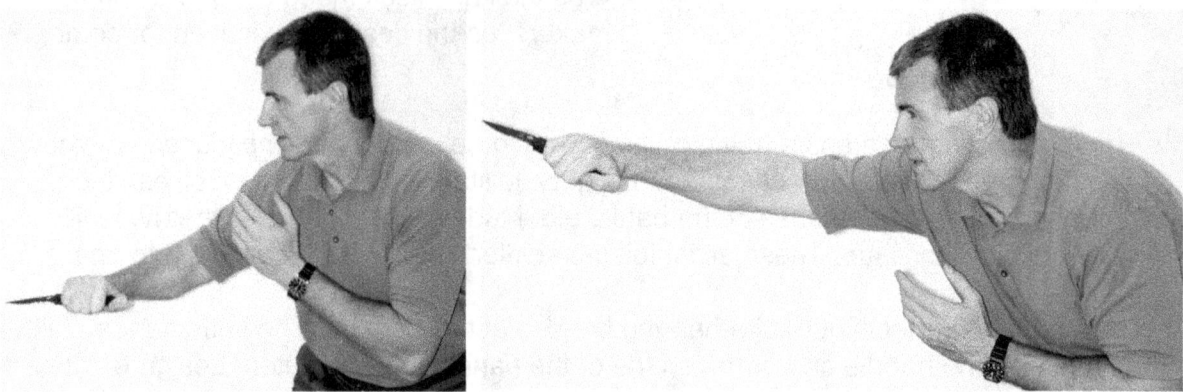

The Myth of the Perfect Knife Grip

All to often, I hear the whisper, "*When you see a guy holding a knife LIKE THIS! Watch out! He really knows what he's doing.*" And by "like this" – he means, a reverse grip

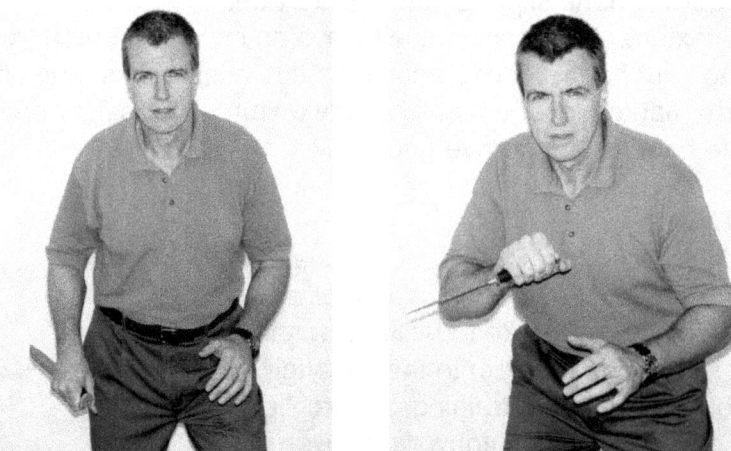

The reverse grip. There is nothing wrong with it. It's just not the one and only perfect knife grip used by real, military "insiders" and properly trained knife practitioners. There is no such thing as the perfect grip. Grips are situational.

This is a myth that seems to perpetuate among the ignorant and it misleads a listener into thinking they have received some secret tip from an insider. Then they pass this on so they too can sound like an insider. I was once working on my laptop in hotel lobby and an onlooker nearby saw the militant subject matter on my screen. He leaned over and informed me that the best knife grip was the reverse grip. He sold car batteries as a profession, but "a Marine told him this once."

There are also some martial arts groups that have their roots rutted in the reverse grip and blindly believe the same advice. But I believe they have never run unbiased experiments in the subject and just mindlessly followed the leader, the pack and the doctrine.

I have no such grip bias. I have conducted observations, studies and experiments with slashing and stabbing with both the saber and reverse group for over two decades. There is no question in my mind that you can do more things with a saber grip than you can with a reverse grip, least of all enjoy a greater range and reach with a saber style.

Simply and scientifically however, the fact is there is no one perfect knife grip, only the best grip for the moment. We fight standing, kneeling, sitting and prone in an unpredictable format of positions and situations. Those moments change. All fights are situational, as are hand grips. Knife practitioners should not force their favorite grips into situations. This is foolish prejudice.

I would prefer a knife combatant be fully versed in both grips and transition to them as needed. For me, I generally prefer a saber grip for most standing problems and a grip for most knife ground fighting problems. Most. Generally. Not always.

You must find such choices out through your own experimentation. Either way, a defender must learn and respect both grips, when, why and where a need or transition to them maximizes their survival and victory.

Two important safety and survival tips have been universally ignored in both military and martial training. Since combat is mobile and forceful, the ice pick/reverse grip has a tendency toward accidental self-injury. A key point overlooked in knife training is the basic concept that the reverse grip knife is pointing back at you and a saber grip is pointing away from you. When you roll? Spring up and finish with slashes and stabs.

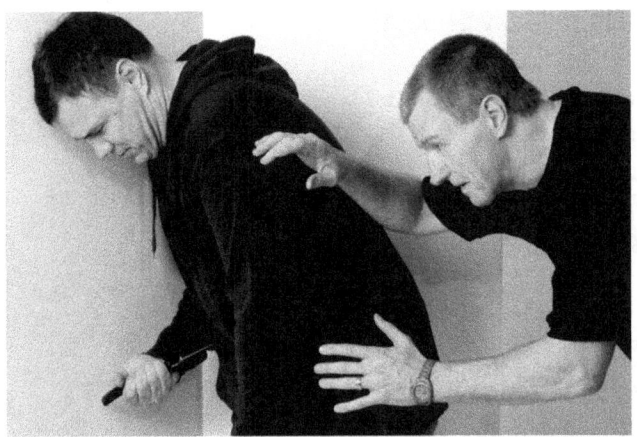

Keep the knife tip pointed away from you. The same problem can be said in combat while standing. Reverse grips can be intentionally or by accident, rammed into the holder's body.

When tripping, falling or forced to roll forward, sideways or backward, a practitioner must be aware of his knife tip, certainly with a reverse grip. Military and criminal justice knife encounters have recorded numerous occasions where knife welders has stabbed themselves by accident when hitting the ground. Practice a roll while holding a knife.

Knife Grips

Suffice to say that a solid grip on your drawn knife is important. One of the things I've written extensively about since the 1990s is what I call the *Cancer Grip*. This very, very silly thing any citizen, soldier, or even a small child would look at and know would not cut a baloney sandwich on soft white bread. The thumb and the ball of the thumb simply MUST be on the knife to stab or slash and keep hold of the knife. Normal people look and say, "Why of course!" But you would be surprised. Not all people agree.

You might receive some knife training from some of these groups, so beware. Usually they are Filipino or Silat based, groups which are automatically presumed to be knife experts. Do not blindly follow everyone and everything.

Probably the most trouble I have gotten into in the established martial arts world was over this very silly knife grip. Worldwide, I was accused of defaming and disrespecting famous Filipino "Godheads." They called me rude and disrespectful. Some martial supply companies refused to sell me gear. A Kali group in Germany, declared that:

"... what Hock doesn't know is that, yes, we hold the knife that way, but when we stab or slash, we grab down on the handle completely during the stab or slash."

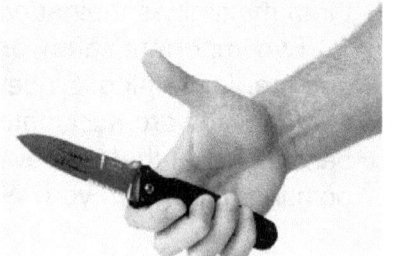

The cancer grip.

So, we are then to presume that if one is unlucky enough to be in a knife duel, or face someone unarmed, or face someone with a stick, you are to half-hold a knife in a ridiculous, useless, unsolid manner while "dueling?: Then you predict an accidental, incidental contact in the chaos of a duel, and...squeeze the knife handle. Then open your hand up again to an incomplete grip? That's your plan? If you see an instantaneous opening from an attack, you quickly squeeze the handle, cut or slash, then retract back to a worthless grip?

There would seem to be no sensible excuse for this. It is more than obvious that any contact with the knife, accidental, incidental or on purpose would cause an easy disarm from that worthless grip.

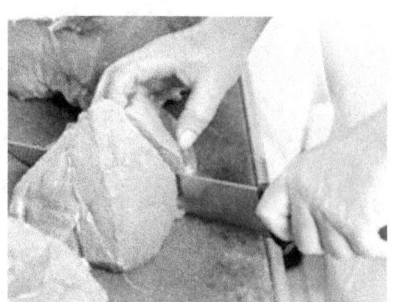

Every butcher uses their thumbs and ball of their thumb to cut meat.

But, there is more, *"Well Hock, that high thumb is used to hook and catch the knife bearing wrist."* So this one hard-to-do, very low probability trick is offered as the

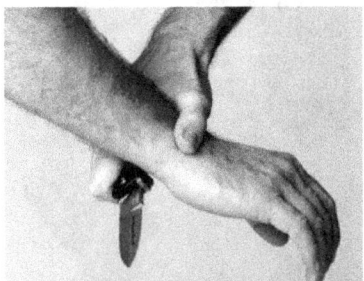

reason we must over-use, over train, see, fight with a terrible stupid, lame grip? A grip with high probability, knife loss upon any contact?

Even just phasing in and out of this lame, artsy, Cancer Grip, as famous instructors mindlessly do when teaching, is a thoughtless and dangerous example.

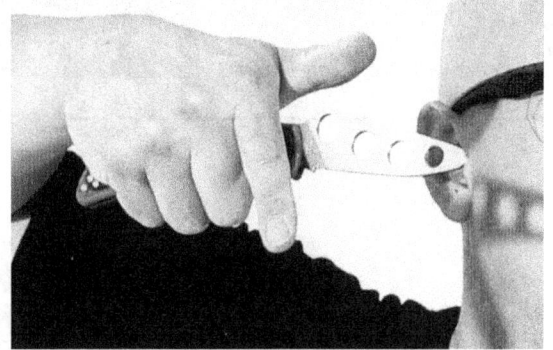

I am told that to this day, you will still see some of the very big FMA names demonstrating and teaching knife drills and slipping off mindlessly and way too often into the lame Cancer Grip. It's prissy and artsy and looks cool? It's an addictive dance. In my travels, I too still see students mindlessly practicing with the worthless Cancer Grip. They are all FMA vets. I began collecting photos of them. I will display just a few of them here, as I avoid showing much of their faces, because I do not want to embarrass them as people. I want you to know not to do these mindless, stupid grips, lest it become your muscle memory. These photos are from magazines, seminar and course ads, webpages and even book covers. These are all instructors you have entrusted with your life. Some are quite famous. Challenge their ideas. One of the worst problems with martial arts is "system worship" and "system-leader" worship where these things creep inside doctrine.

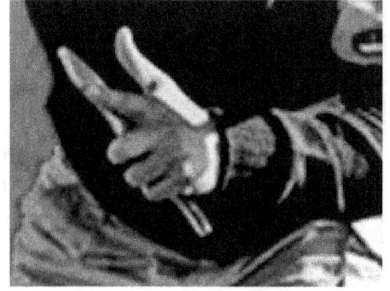

Military and crime forensics report that stabs are more successful than slashes. You cannot put your finger out on the blade. It will hamper your stab, as the stab will only penetrate up to your finger, limiting your depth success. Why limit your success by limiting the penetration of the stab with your finger?

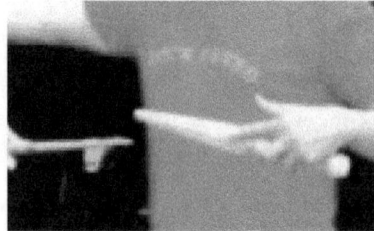  

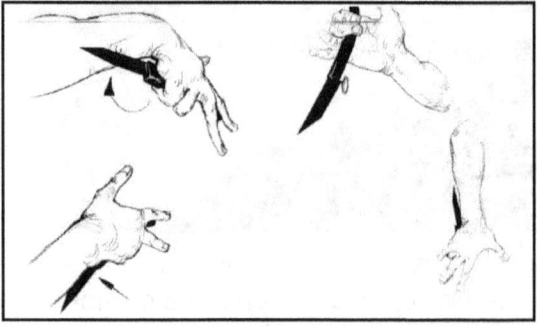 

This is some instructional artwork for knife manipulation. I will not embarrass the system head or the system by naming them. It is all open-finger handicapped, stupid, artsy junk.

WTF?

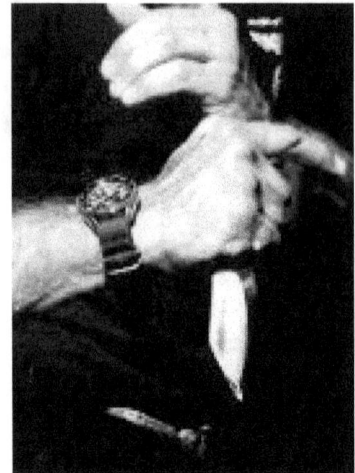 

More "famous" people" letting themselves be photograhed in this manner in promos and covers. I have collected about a hundred of such photos, but you get the "picture" with these samples.

The knife is a great equalizer, but not God's gift to equalization. As a result of this equalization factor, you can do a lot of stupid stuff and maybe get away with it. If you want to maximize your chances for survival, stop this infection. Use your hand and all your fingers to hold your knife, to retain your knife, whether presented or slashing or stabbing. Military veterans laugh at such artsy nonsense.

The Knife Grip Lanyards

When gripping the knife, a lanyard is an option many veterans prefer. This offers greater grip support and weapon retention. The lanyard itself may be made from any stout cord or rope. A common lanyard choice is paracord.

Lanyard Length. The lanyard must be measured to fit the size of your hand. The two common length choices are short (one wrap around hand), or long (perhaps two wraps). People hook their thumb, the hand between the thumb and the "pointy" finger, or the wrist. The length is important and should be established by trial error. Lanyards can be important. People lose their knifes in use – be it in the-backyard, camping or combat. In a worst-case scenario, lanyards (and their big-brothers like long gun slings) may be used by the enemy as a pulling device against your balance, and even for a takedown. If the cord is long enough, the soldier can release the lanyard from his or her thumb or grip, narrowing their hands in a curving, cupping motion and let the lanyard slip off, rather than their whole bodies being tossed on the ground in a takedown. This counter takedown release has been taught in military knife courses.

Lanyard Optional. Many people dislike knife lanyards. In sudden combat you may not have a chance to apply the lanyard to your hand. If this is the case? Where will this lanyard be in your common daily carry? Your desperate quick draw? Will the loop catch on items in your environment? To lanyard or not to lanyard? The choice is yours.

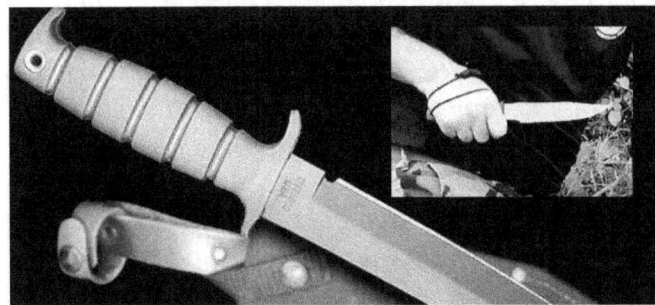

Many knives come with a lanyard upon purchase, or at the very least, come manufactured with a hole in the handle for lanyard application.

French Foreign Legionnaire veteran and global military combatives expert Nick Hughes demonstrates the lanyard with knife grip.

Black Box Knife Combat Files

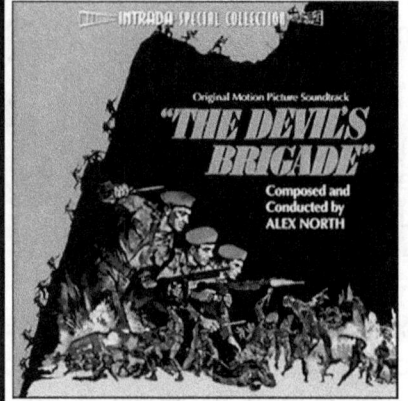

"It was not uncommon for some military knife grips to allow for a thumb positioned beyond the hilt/guard. The Devils' Brigade: The Original Special Forces Knife – By Danny M. Johnson, of the U.S. Army Research Center

"U.S. Army Special Forces units today trace their lineage and honors directly to the First Special Service Force (FSSF), the joint Canadian-American organization formed in July 1942 at Fort William Henry Harrison, Montana. This regiment was an all-volunteer outfit with exceptionally high standards and extensive and rigorous military and physical training. The "Forcemen" (as they called themselves) were trained in airborne operations, mountaineering, amphibious operations, and ski tactics.

The FSSF fought under Allied command with great bravery and considerable success. The Force got its nickname, "The Devil's Brigade," during the Italian campaign from a passage in the captured diary of a dead German officer who had written: "The black devils are all around us every time we come into line and we never hear them."

The Forcemen were issued special gear including specialized weapons, leather jackets, mountain climbing gear, parkas and other cold weather equipment. As an elite "commando type unit" that would operate near or behind enemy lines. The knife envisioned by its commander, LTC (later MG) Robert T. Frederick, was actually a group effort among his staff for the perfect fighting knife. FSSF close quarters combat instructor, Dermot Michael "Pat" O'Neill, a former sergeant in the Shanghai Municipal Police Force, suggested the blade profile. O'Neill gave great thought to the needs of these special troops as it related to close quarters combat.

The V-42 is a legendary knife that today is an enduring symbol of the most elite of the U.S. Army. No article of equipment is as synonymous with the First Special Service as their V-42 Fighting Knife. The V-42 was featured a 7 5/16 inch blued stiletto blade, and had an overall length of 12 inches The handle was made of finely serrated leather washers and the pommel ended in a short point for use as a "skull crusher." On the early V-42s, this skull crusher point was quite sharp and caused problems by catching on clothing, forcing soldiers to file it down. The knife was marked "CASE" on the ricasso, and a section of the rear blade near the blade shoulder was serrated for use as a thumb rest to help align the blade during a thrust."

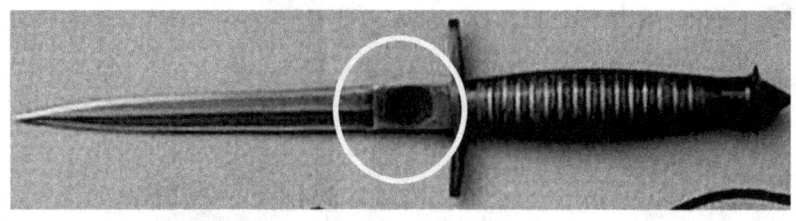

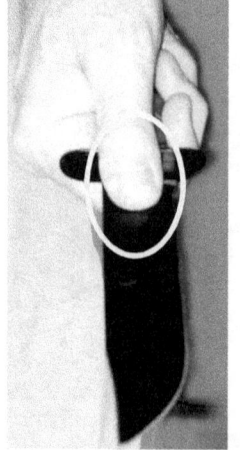

CHAPTER 8: KNIFE STRESS QUICK DRAWS

The best quick draw is one when you pulled out your knife *before* you actually needed it. A knife is worth little or nothing if the soldier cannot access and deploy it, especially under combat stress. Historically, this quick-draw/fast-deployment tactic is the least practiced and under-emphasized movement in military or martial arts knife training, or surprisingly even in any survival training.

Since the 1990's, new to the battlefield and the street is the wide proliferation of the modern tactical folding knife, creating even more knife-opening, deployment problems.

A soldier who conducts checkpoint, or occupied territory, or force protection interviews and actions needs to know the methods on how a civilian, a terrorist or a disguised soldier might draw weapons, concealed or overt. Or the soldier, having lost other ammunition and/or other weapons in his arsenal, or perhaps on a covert assignment, may need to suddenly draw a fixed-blade knife or a folding knife. Enforcement and citizens need these skills.

This section contains demonstrations of carry-sites, methods and gear issues concerning the stress, combat quick draw of a military tactical folder or fixed blade knife.

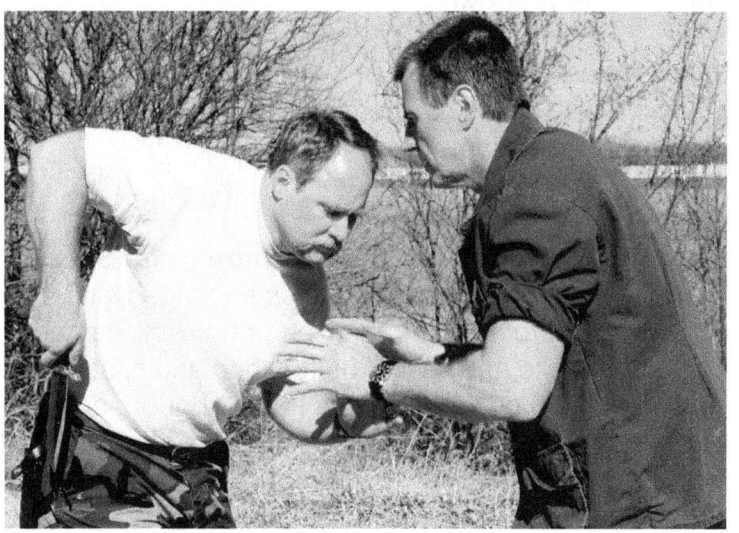

In the matrix of quick draws, you are dealing with issues in:

- weapons.
- sheaths, clips, lanyards.
- carry sites.
- methods of concealment.
- hand moves to carry sites.
- distractions and fakes.
- opening up folders.
- right side carry.
- left side carry.
- right hand to right side pull.
- right hand to left side pull.
- left hand to left side pull.
- left hand to right side pull.
- double edge knife.
- single edge knife.
- single edge with half edge.
- edge in? edge out?
- saber grip.
- reverse grip.
- factory sheath.
- custom sheath.
- ambidextrous sheath.

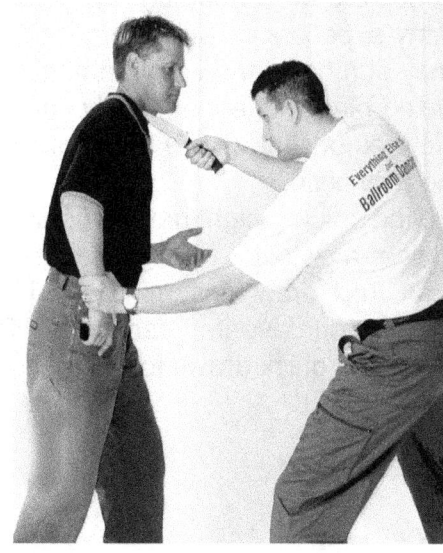

Weapon Carry Sites and their Relation to Quick Draws

Primary Quick Draw Sites
These are carry sites that allow for the fastest quick draw. Usually they are around the belt line, in and around the pockets of pants and jackets, at times in a shoulder holster type rig. Also, consider the inventiveness of the enemy.

Secondary Quick Draw Carry Sites
These are carry sites where the weapon is somewhat buried under clothing. Usually they are a neck knife rig inside shirts or jackets, a boot knife rig. In the armpit inside a bullet-proof vest. In a compartment pocket of a tactical vest.

Tertiary Quick Draw Sites
These are classified as lunge and reach, or also hidden weapons categories. The enemy might pass a visual or physical inspection, yet may lunge for a previously concealed weapon in a pre-meditated plan or a new improvised one. The weapon he lunges for? It may also be your own.

Learn the on-body and off-body sites of you and your enemy. Learn and exercise efficient quick draws. Learn and exercise early-phase, mid-phase, and late-phase counters to weapon quick draws.

Issues in Quick Draws: The Common Knife Sheath

A military issue knife usually comes packaged with a custom fit sheath made from leather or any number of new durable soft or hard substances. Once this sheath is secured on the belt or carry site, the design of the knife and custom sheath limits the soldier to a certain kind of weapon pull. Unless you have a double-edged, basic commando design knife (the suggested best) the long sharp edge of the knife will either be facing in or facing out as many issued military knives are not double-edge, commando style.

    A soldier may practice with a certain knife grip in basic and advanced training schools. For example any number of military schools will supply trainees with rubber knives that are usually double-edge, based on my experience teaching all four branches of the U.S. Military and my experiences with international services. But, then in the field, they may be issued, or may purchase and carry another design knife such as a single-edge, tactical folder, or a fixed-blade knife that is fully sharp on one edge and half sharp on the other. Suddenly the methods they were taught in schools do not exactly apply to design and sheath of the new knife.

    In training, the soldier might have a right-side, belted sheath. But, in the battlefield the soldier may wear a pistol on the right side, now forcing the knife carry to the left side. And now the stress quick draw must be done with the left hand and the custom-shaped sheath may not allow for your favorite, practiced knife grip. Considering the militaries of the world have ignored stress quick draws for a few centuries, it is no wonder this is a modern problem.

In an abstract way, this is teaching troops to use an M-16 in training and then handing them a bolt-action carbine on the battlefield. It is also like teaching pistol craft and then ignoring the pistol quick draw from a holster. Many factors are the same and many not.

The solution is to either warn the soldier about these issues and teach them the methods of customization, or train them on knives and gear they will use from the school to the battlefield. The latter being almost impossible to enforce, it is better to brief the trainee in the realities of knife and carry options.

 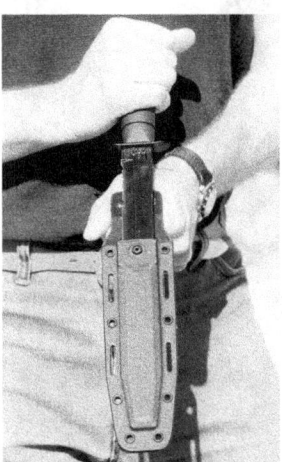 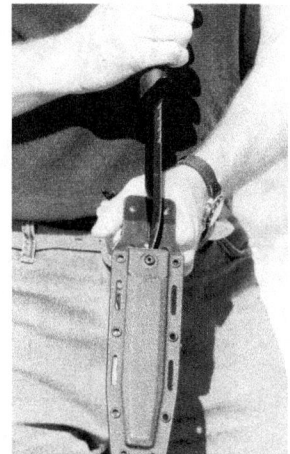 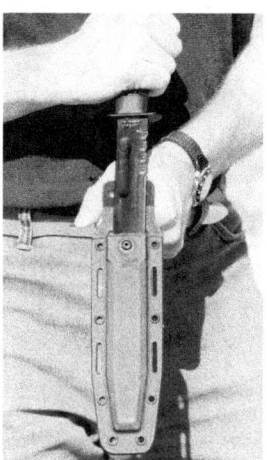

The classic issued and sold sheath. Limiting.

The purchase of the ambidextrous sheath allows for a fixed blade knife to slip in and out, and will allow a soldier's potential, preferred grip outcome.

The Forearm Carry Site. This allows for fast quick draw providing the sleeves are wide and open enough for the hand entry and access to get the handle of the knife. These forearm rigs are improvised or purchased usually in scuba and diving gear stores.

Defending citizens, police and soldiers need to look for wide sleeves and "forearm-scratching" hand movements. Attacking soldiers need to conceal their wide sleeves and their quick draw hand motions.

The Cross Draw. The arm moves to the armpit, or the area between the arm pit and the belt line, or to the belt line, facilitated by shoulder holsters and belt line holsters.

The Napoleon Draw. This allows for fast quick draw providing the shirt or jacket is untucked and/or open enough for the hand entry and access to get to the handle of the knife and pull it from the sheath – a downward yank of several inches. These rigs are purchased almost anywhere.

Defending citizens, police and soldiers need to look for partial open and untucked shirts and "chest-scratching" hand movements. Attacking soldiers need to conceal these shirt issues and sneak their quick draw hand into or up under their shirts.

Pocket Folder Quick Draws. Most citizens, police, security and soldiers now have their own folding knives and have clips on the handle. In fact, all tactical-based folders have these clips so as to hook on vest, belts, harnesses and pockets. The folder has to be accessed and opened. Once the folder is locked open, it takes on the characteristics of a fixed blade knife. Defenders must:
   - define the carry site, pocket or otherwise.
   - define the type of grip possible from the draw.
   - define what hand draws and opens the knife.

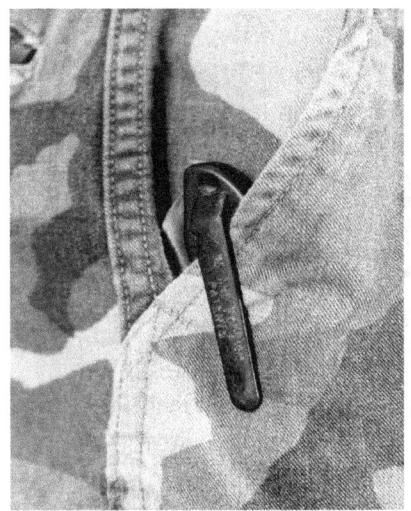

When drawing, do not squeeze the knife clip and knife on your pocket. It will slow the draw.

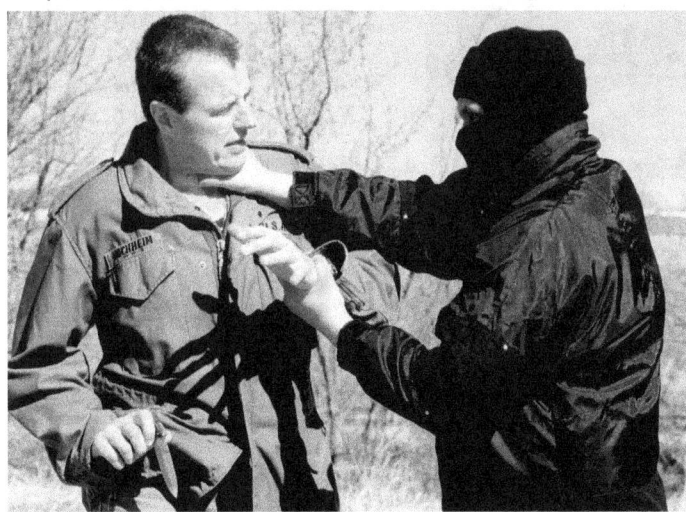

The defender must not squeeze the clip onto the knife handle with his fingers when first grabbing the pocketed or clipped knife. This pinches the knife clip onto the pocket and slows the draw process. The soldier must access the knife with his fingers and lift it from the clipped position minus this squeeze on the clip. The soldier must practice the draw from:
- standing
- kneeling
- seated
- ground positions
- under striking assaults
- under grappling assaults
- all of the above under the increasing stress attack of a trainer.

(Gloves! In many parts of the world, winter is coming or already there! People wear gloves and usually "normal store-bought" gloves. These are not completely conducive to drawing knives (or guns) and are not tactile enough to fully feel open a folder in comparison to the naked hand. Tactical minded people obtain the best gloves for these stressful jobs.)

The Lanyard Sheath Pocket Draw

 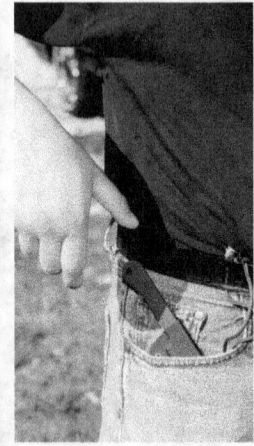 

The lanyard pocket draw. The lanyard is looped through the bottom of the sheath and the belt. The knife and sheath is pushed into the pocket. The enemy reaches into his pocket and grabs the handle, pulling the whole knife and sheath with it, out.

In the course of presenting the knife for combat, the lanyard hits its length limit, the sheath is pulled from the knife and the knife is drawn.

Soldiers, security, police and civilians should be watchful for such strings, cord and rope from the belt line down into the pocket.

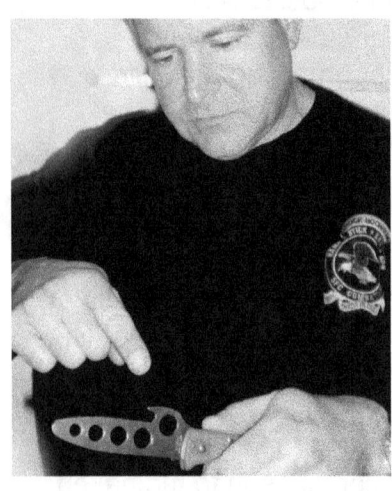 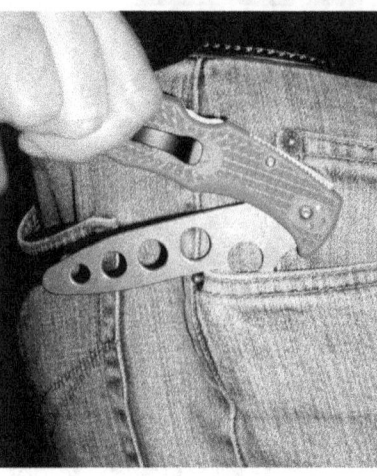

Blade Cut-outs and Hooks. Many folders are made or cutomized with belt line catches that hook the pocket line of pants and open the blade as it is extracted from the pocket. U.K.s Joe Hubbard shows this "hook" on an Emerson trainer.

The Boot Carry: "The Shoe String Draw." An ankle sheath, or "boot knife" is a common concealed, secondary carry site. And for some, a primary carry site. This weapon's pull is often facilitated by a ruse to tie a loose shoe string. It has been used after tripping, falling and in ground fighting. Be leery of any suspicious person suddenly dropping to tie their shoelace.

The Gypsy Draw. Backpacks often have knives and machetes secured to them in some way, top, sides or bottom. I have witnessed knives hooked onto the military back packs of soldiers in such ways that the soldier could not even reach them, yet close and/or captured enemies may reach that knife.

The Spine Carry. This shoulder holster rig carries a handle-down fixed blade knife along the backbone.

The Use of Distraction Quick Draw. Always try to use deception and trickery with every tactic to confuse the enemy. This holds true with stress combat quick draw. Be leery of any suspicious person holding any items or near any items they can lunge for and throw.

The "Palm Stick" Failsafe Option. If a defender is unable to open a folder in the middle of a stress draw, the soldier should use the weapon as small club or, as some refer to it, as a "palm stick." The folder must first be chosen with the option in mind. How well does the folded folder fit in your hand? If your struck someone while holding it, would this hurt your hand? Can youuse the knife as a "brass knuckle" device? Will any edges of the closed folder hurt your hand when striking? Do both ends protrude beyond your hand enough to strike with?

Exercise Drills for Stress Knife Quick Draw
Select the carry site and weapon, then execute:

    Drill 1: Stress knife quick draws while wind sprinting.
    Drill 2: Stress knife quick draws while running an obstacle course.
    Drill 3: Stress knife quick draws while stepping in Combat Clock Ground patterns.
    Drill 4: Stress quick draws while ground fighting.
    Drill 5: Use the **Force Necessary**, **Stop 6** Rattlesnake Series to practice stress quick draws with a partner (see upcoming chapters).

        Stop 1: Stopped at "Showdown" standoff distance – draw and take appropriate action.

        Stop 2: Stopped at hands – capture the enemy's weapon hand, draw and take appropriate action.

        Stop 3: Stopped at forearms – draw and take appropriate action.

        Stop 4: Stopped with hands at neck/chest/biceps line – draw and take appropriate action.

        Stop 5: Stopped at bear hug/clinch – draw and take appropriate action.

        Stop 6: Stopped on the ground – draw and take appropriate action.

Samples of stress combat quick draw combat scenarios will appear later in this this book.

CHAPTER 9: THE KNIFE COMBAT CLOCK

The clock points. I was familiar with military terminology and concept from my Army experience. If you were on a foot patrol and the point man suddenly shouted, "enemy at 2 o'clock!" Everyone would instantly look in that direction. The same for pilots – who also have both a vertical clock and a horizontal clock. "12 o'clock high!" Simple. Quick. Effective. Unforgettable.

Yet, scores of differing police and martial arts training systems are not clock-based. Their elaborate weapons angles of attack were disjointed and forgettable, a major problem in this frustrating rat race of systemalogies, and the various lines of attack protocols each used for hand, stick, knife and gun tactics. Each try to outsmart or out-do each other, rather than focus in on maximized education. The worst in my opinion are the two extremes – the over-simplified and the over-complicated. I began to ask myself, how are all these directions of combat the same? It became clear that attacks universally come in from the center, high or low, or right or left sides, whether standing or on the ground.

No matter the weapon, the angles/directions are the same. I returned with trust to the simple military "combat" clock. The clock face is an imprinted image in our minds since early childhood. The simple angle of attack pattern is right on your wrist, work or play. I discovered, or better said, I re-discovered the simple, military clock method as a training foundation. Stand it up or lay it down, you have an unforgettable pattern to teach, memorize and work from.

Basic Hand, Stick, Knife, Gun Combat Clock Training:
    12 o'clock from axis to above.
    3 o'clock from axis to the right.
    6 o'clock from axis to below.
    9 o'clock from axis to the left.
    Axis point is the center.

Advanced Hand, Stick, Knife, Gun Combat Clock Training:
    From the axis center point out to all 12 numbers of the clock.
    This offers more precision training if needed.

The Combat Clock is Used to:
    Learn knife manipulation and solo command and mastery skills.
    Maneuvering/footwork - organize attack and defense footwork if laid horizontal.
    Target Spotting direct fire and locate enemies with a vertical and horizontal clock.
    Delivery system - use to deliver angles of attack.
    Organize attack striking, hooking/slashing strikes if set vertically.
    Organize attack shooting/stabbing/thrusting points if set vertically.
    Organize defensive moves if set vertically.
    The clock as a clock. Coordinate mission timing.

Basic Combat Clock Training: 12, 3, 6, 9, Axis

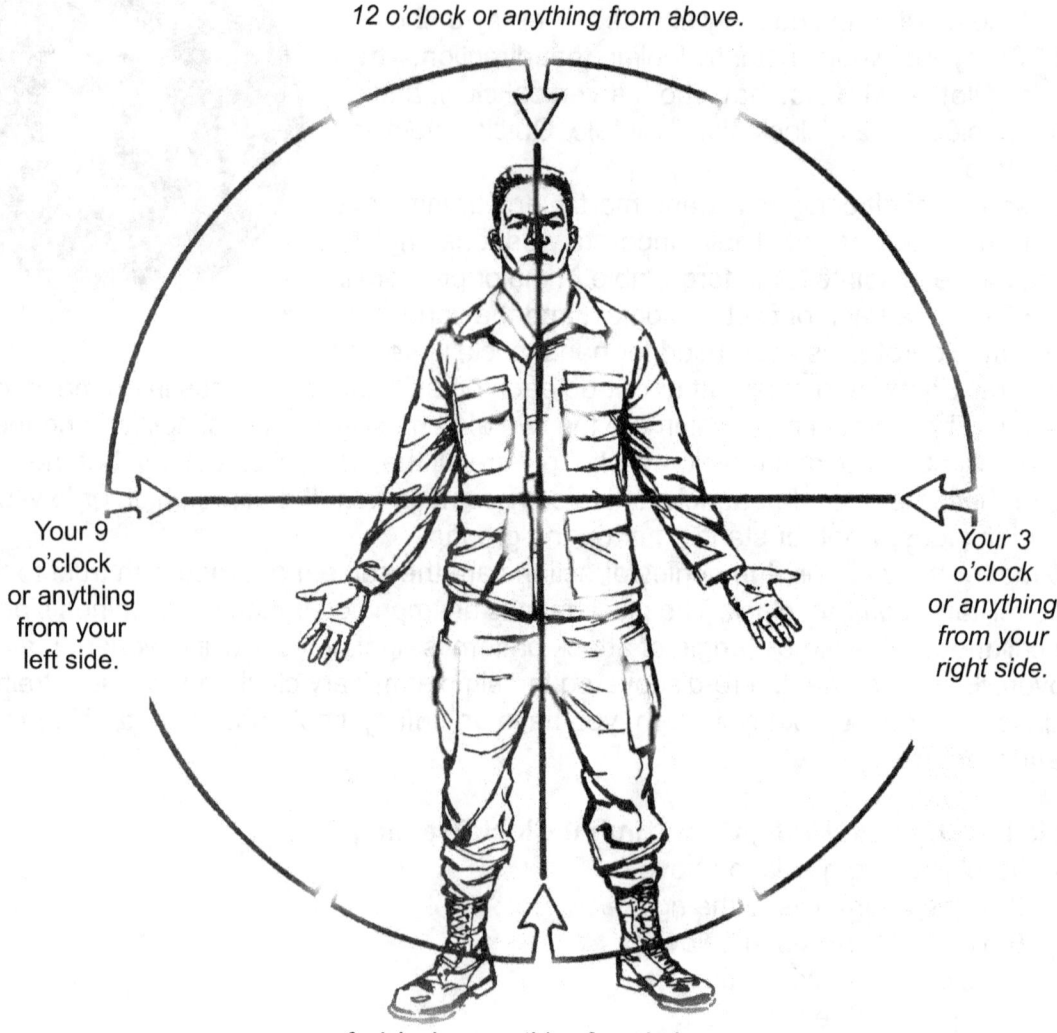

In this Basic Training Format using the four corners and the center of the clock, an instructor can have trainees operating in class and interacting with others quickly and efficiently. As detailed in a subsequent section and summarized here, a trainee will:

Execute footwork patterns on the horizontal clock.
On the vertical clock, execute the following:
. Knife saber stab at the 12, 3, 6, 9 and axis.
. Knife reverse grip stab at 12, 3, 6, 9 and axis.
. Knife saber slash at the 12, 3, 6, 9 and axis.
. Knife reverse grip slash at the 12, 3, 6, 9 and axis.
. Block at 12, 3, 6, 9 and axis.

Advanced Combat Clock Training: 1 o clock through 12 o clock and Axis

This Advanced Training Format uses more specific targeting. By targeting the complete 12 numbers of the clock and the center of the clock, an instructor can have trainees operating in class and interacting with others. As detailed in a subsequent section and summarized here, a trainee will:

    Execute footwork patterns on the horizontal clock.
    On the vertical clock, execute the following:
        . Knife saber stab at the 1 through 12 points and axis.
        . Knife reverse grip stab at the 1 through 12 points and axis.
        . Knife saber slash at the 1 through 12 points and axis.
        . Knife reverse grip slash at the 1 through 12 points and axis.
        . Block at the 1 through 12 points and axis.

The Combat Clock as an Attack or Defense Model

The window of combat, as it pertains to the standing, ready, fighting positions, is where you maintain your limbs as a base to move, strike, block and return to. The window of combat is the combat clock encircling you and loosely bordered by about your shoulders down to your mid-thighs. This "window concept" is especially important in knife vs knife dueling, a subject covered in depth in another chapter of this training manual.

Imagine positioning your empty hand at various clock numbers, or your knife at various clock numbers. This is a method of training that has deep potential and runs from the very simple to the very complex should the training require advanced training such as ambush and assassination criteria. Using the Combat Clock as an attack or defense module the instructor can organize groups into specific knife positions and hand positions.

The Combat Clock as an Attack or Defense Model. Knife at the 4 o'clock line, hand at 2 o'clock.

Militaries from common knife cultures allow their soldiers a variety of stances and flexibility to improvise. The fighting position you choose will be strategically selected to counter the enemy you face before you and one that will fit into the situation. There is no such thing as one perfect knife fighting stance. Several of the militaries of the world now mandate a knife back fighting stance for all occasions, and such a broad mandate is a serious mistake. The writers of such doctrine are inexperienced and inept and are not maximizing the lives and safety of their troops.

The Combat Clock used as a Attack Defense Model, as applied to ground fighting positions and the position of your enemy.

The Combat Clock for Knife Attack, Knife Defense and Combat Footwork

Note 1: In the last 4 decades I have taught thousands and thousands of people from utter novices to experts, cadets and rookie cops to martial arts black belts, from all over the world, and I can get them to interact with each other in mere moments by using this simple basic clock point format.

Note 2: Remember, there is a difference between target spotting and angles of delivery. When practicing these clock angles alone and "in the air" so to speak, you are only learning "weapon manipulation" skills. Whatever the weapon, be it a hook punch or a stick strike - "Solo Command and Mastery" skills - as I call them, DO NOT assign strikes to body part targets in your official nomenclature.

Do not always call a 12 o'clock strike a "head strike," or a 3 o'clock hit a "heart strike." In combat we do not know what position the enemy's body will be in. Plus, your first strike, stab or shot will probably change the template! You are only practicing a directional delivery skills when working alone.

If you always imagine hitting specific body parts in your solo workouts, then change up the targets in your mind. Keep your mind open to options, to hunting moving targets that change. A downward 12 o'clock strike might be the face, or to the top of an arm thrusting a punch or stab at you.

Note 3: Sure, we see digital clocks here and there, but we will see the clock face for another century, maybe longer.

Note 4: I recall in the 1990s, a police academy instructor, kicked back in his office chair, feet on his desk, complaining to me, *"you can't teach these people anything. If you teach them three angles of attack? In six months, they will forget two of them."* Instead of complaining, he should have been judiciously working ways to develop and mold doctrine into unforgettable, high retention methods. Lazy, uninspired bastard. He was wearing a wristwatch, by the way. The combat clock. Simple. Timeless. Unforgettable. Versatile.

Black Box Knife Combat Files

South Pacific, WW II " ... Lee grinned in spite of himself, then started with revulsion. Behind the sleeping guard a dead American soldier had been strung up and bayoneted countless times, like a practice dummy. It looked to Lee like the B company acting C.O., a Lt. Braun. Lee turned away, sick at the sight, realizing the same would be in store for him soon. He struggled to reach the knife he kept inside his boot. After a half-hour of contortions worthy of Houdini, he managed to extract the knife and saw through the thin hemp rope holding his wrists. He crawled behind the sleeping guard and stabbed him efficiently in the back, angling his knife so that it would enter the heart and holding his other hand over the man's mouth."

– Anthony Arthur, Bushmasters, America's *Jungle Warriors of World War II*, St. Martin's Press

CHAPTER 10: THE KNIFE COMBATIVES READY POSITIONS

Knife fighting stances and poses have been taught since sword fighting antiquity, and based more or less on equal-sized weapon duel. But modern, edged-weapon combat is more about balance and power in motion than posing in particular fighting "stances."

Poses are truly best used for:
    a) to line up and uniformly organize soldiers into positions for training classes.
    b) at the start of some fights if sufficient distance in involved.
    c) to intimidate the enemy into surrender or flight.

A ready position, is akin to a sports position, Balance and power ready. You will transition though the standing, kneeling and ground positions. I like to instruction 9 action-ready positions, and one neutral position.

This neutral one can be, as the famous Ed Parker of Kenpo Karate use to call it, the oblivious "bus stop" stance, where you are totally oblivious to your surroundings. Daydreaming. Or the hands down, "unready" stance could be a bluff to an attacker. So that bus stop/neutral stance can be two-fold.

Defenders must familiarize themselves with more depth and issues in the neutral stance. This position can be more of a stationary stance and be a deceptive starting point for attack and counter-attack.

A soldier dispatched on an open or covert assignment might become suspicious of a confrontation and, like the plainclothes soldier on the left in the photo, he may secretly draw his knife and hold it concealed from an approaching or confrontational enemy. The knife may by in a saber or ice pick grip.

Otherwise, knife combat is about balance and power in motion.

The "bus stop" stance. Are you daydreaming, or ready-and-bluffing? Is he?

Standing around unarmed? Or hiding a knife?

Concealed saber or concealed reverse grip.

To start these motion studies here, we will review the past through the evolution of the ready positions and situations of knife combat. Be ready to fight from any one of these positions.

Ready and/or Fighting Position 1: Knife Forward.
The knife leads the way. The knife-bearing limb is forward. The knife-side leg is forward. The body is in an athletic position.

Ready and/or Fighting Position 2: Knife Neutral.
The knife-bearing limb is not forward or back, but in a neutral position. Either leg could be somewhat forward. Knife neutral is also used in ambush positions, as shown in detail later.

Ready and/or Fighting Position 3: Knife Back.
The knife-bearing limb is back. The knife-side leg is back. The empty hand is up front. The body is in an athletic position.

Ready and/or Fighting Position 4: Knee High.
The defender is one knee high or down on both knees. Defenders include seated training in this category. This includes all top-side fighting positions of a ground fight. You fight people standing over your, your height, and below you. This is the top-side of a knife ground fight.

Ready and/or Fighting Position 5: On your back.

Ready and/or Fighting Position 6: On your right side.

Ready and/or Fighting Position 7: On your left side.

1: Knife Forward

2: Knife Neutral

3: Knife Back

Position Practice Set 1:
- Transition through all of these positions in a flowing series.

Notes:
Point 1: The knife can be in either a saber grip or ice pick, reverse grip.

Point 2: The empty, support hand should be up in the window, that clock circle of combat ready to take action.

4: Knee High

6: On Your Sides

5: On Your Back

CHAPTER 11: KNIFE ATTACK POINTS

The Four Knife Strike Points

The knife has four striking points in close quarter combat, the tip, the edges, the flat of the blade, and the pommel. Exotic design knife might have a few more unusual features for striking – such as an odd, large-shaped hilt, but most practical knives do not have such features. A folding knife has the same four key strike points. A properly trained combatant knows on whom to use these points, where, when, how and why

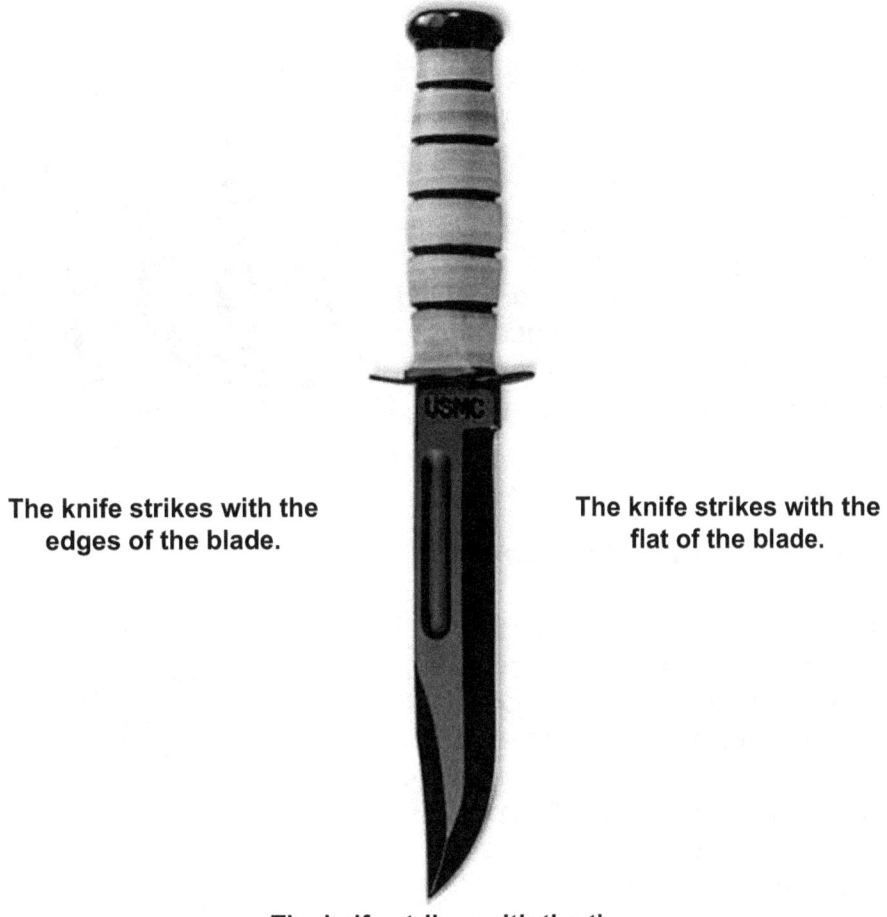

The knife strikes with the pommel.

The knife strikes with the edges of the blade.

The knife strikes with the flat of the blade.

The knife strikes with the tip.

The knife strikes with:
- Strike 1: The tip.
- Strike 2: The edges of the blade.
- Strike 3: The flat of the blade.
- Strike 4: The pommel.

The 4 Ways the Knife Arm Attacks

The knife may be held in a saber or reverse grip, and it may be delivered on all angles of a combat clock, but a soldier's arm can only deliver a knife attack four fundamental ways. With a power lunge, a pump, a hook or a thrust. In the science of knife/counter-knife tactics, understanding these four arm deliveries is an essential concept of doctrine. These also apply to hand strikes, kicks and impact weapon attacks.

The Big 4 Attacks!
You MUST remember these.
1: A power lunge.
2: Hit and retract.
   That is either a...
3: Hooking attack,
4: Thrusting attack.

Arm Delivery 1: The Power Lunge Arm

A knife is often a weapon of desperation and anger, and this frequently, physically translates to a dedicated power lunge that buries deeply into the enemy, or if missed, the lunge has a committed and considerable full-body, follow-through. This follow-through can be advantageous to the dodging defender. The arm lingers in his range longer for counter-attacks, grabs, etc. The attacker may be off balance. The lunge attacks on all angles of the Combat Clock. A hook or a thrust attack.

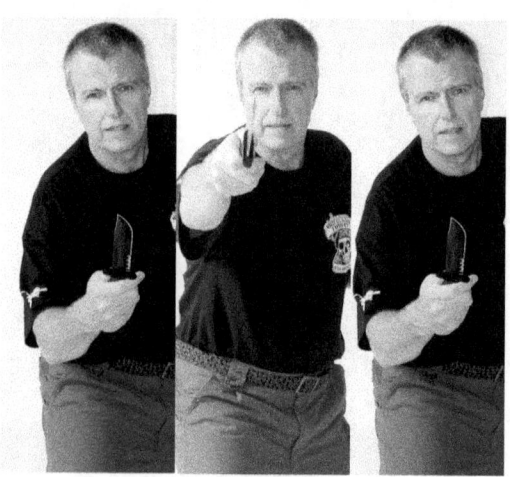

Arm Delivery 2: Hit and Retract, The Pumping Arm

The pumping knife does not linger and does not deep lunge/power drive. Instead, like a boxer's jab, the knife pumps in and back out. The arm does not linger in counter-attack range. The weapon bearing limb is harder to grab because of this spring action. The attacker's body is more balanced. The pumping, jab attacks at all angles of the Combat Clock. This too is either a hook or a thrust.

Arm Delivery 3: The Thrusting Arm

The knife is delivered in a straight line thrust to its target. The thrust begins from the original knife position to the target. The thrusting arm attacks at all angles of the Combat Clock.

Arm Delivery 4: The Hooking Arm

The knife is delivered in a hooking, curving line of attack to its target.
The hooking arm attacks at all angles of the Combat Clock.

CHAPTER 12: FOOTWORK AND MANEUVERING

There are basic principles of movement that the knife fighter must know and apply to successfully defeat an opponent. The principles mentioned are only a few of the basic guidelines that are essential knowledge for knife combat. There are many others, which through years of study become intuitive to a highly skilled fighter.

1: *Physical Balance.* Balance refers to the ability to maintain equilibrium and to remain in a stable, upright position. A fighter must maintain his or her balance both to defend himself and to launch an effective attack. Without balance, the fighter has no stability with which to defend, nor does the defender have a base of power for an attack. The fighter must understand two aspects of balance in a struggle:

> *a)* How to move his or her body to keep or regain his or her own balance. develops balance through experience, but usually the defender keeps his feet about shoulder-width apart and his knees flexed. He lowers his center of gravity to increase stability.

> *b)* How to exploit weaknesses in his opponent's balance. Experience also gives the hand-to-hand fighter a sense of how to move his body in a fight to maintain his balance while exposing the enemy's weak points.

2: *Mental Balance.* The successful defender must also maintain a mental balance. He must not allow fear or anger to overcome his ability to concentrate or to react instinctively in hand-to-hand combat.

3: *Position.* Position refers to the location of the fighter (defender) in relation to his opponent. A vital principle when being attacked is for the defender to move his body to a safe position-that is, where the attack cannot continue unless the enemy moves his whole body. To position for a counterattack, a fighter should move his whole body off the opponent's line of attack. Then, the opponent has to change his position to continue the attack. It is usually safe to move off the line of attack at a 45-degree angle, either toward the opponent or away from him, whichever is appropriate. This position affords the fighter safety and allows him to exploit weaknesses in the enemy's counterattack position. Movement to an advantageous position requires accurate timing and distance perception.

4: *Timing.* A fighter must be able to perceive the best time to move to an advantageous position in an attack. If he moves too soon, the enemy will anticipate his movement and adjust the attack. If the fighter moves too late, the enemy will strike him. Similarly, the fighter must launch his attack or counterattack at the critical instant when the opponent is the most vulnerable.

5: *Distance.* Distance is the relative distance between the positions of opponents. A defender positions ones self where distance is to his advantage. The hand-to-hand fighter must adjust his distance by changing position and developing attacks or counterattacks. He does this according to the range at which he and his opponent are engaged.

6: *Momentum.* Momentum is the tendency of a body in motion to continue in the direction of motion unless acted on by another force. Body mass in motion devel ops momentum. The greater the body mass or speed of movement, the greater the momentum. Therefore, a fighter must understand the effects of this principle and apply it to his advantage.

    a) The fighter can use his opponent's momentum to his advantage-that is, he can place the opponent in a vulnerable position by using his momentum against him.

        i. The opponent's balance can be taken away by using his own momentum.

        ii. The opponent can be forced to extend farther than he expected, causing him to stop and change his direction of motion to continue his attack.

        iii. An opponent's momentum can be used to add power to a fighter's own attack or counterattack by combining body masses in motion.

    b) The fighter must be aware that the enemy can also take advantage of he principle of momentum. Therefore, the fighter must avoid placing himself in an awkward or vulnerable position, and he must not allow himself to extend too far.

7: *Leverage.* A fighter uses leverage in hand-to-hand combat by using the natural movement of his body to place his opponent in a position of unnatural movement The fighter uses his body or parts of his body to create a natural mechanical advantage over parts of the enemy's body.

8: *Footwork.* Combat footwork is a mixture between sport footwork, as in kick boxing, and obstacle course work. The two realms together create the mobile combatant.

    a) Stationary right foot – left foot steps forward and back: This is an excellent advance and retreat step that keeps you in range. From a fighting stance, leave your right foot stationary and step forward with your left foot, then step back.

        i. Combat Clock sample - right foot stays on clock axis. Left foot moves from 10 to 2:

b) Stationary left foot – right foot steps forward and back: This is an excellent advance and retreat footwork that keeps you in close range. Do this from a stationary left foot, with the right foot stepping forward and back. This is advance and retreat footwork.

    i. Combat Clock sample – left foot stays on clock axis. Right foot moves from 2 to 5.

 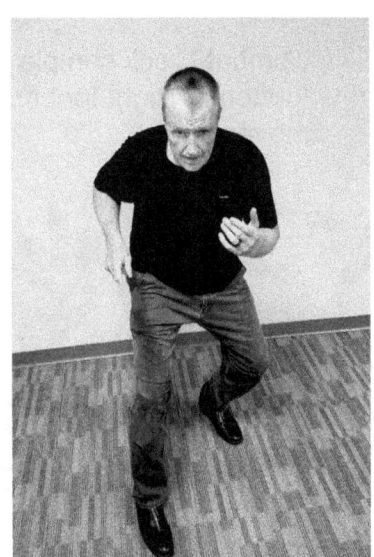 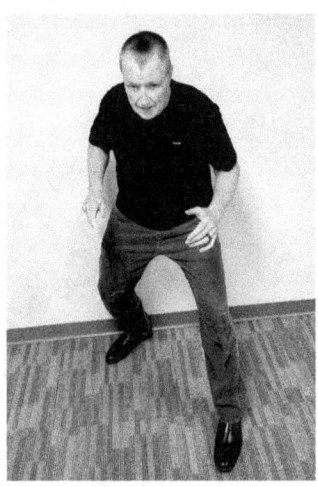

An example of this in and out footwork. Right foot steps in. Right foot steps back.

    c) Shuffle footwork (the pendulum): If you are shuffling forward, let your rear foot come forward near your front foot, displace it, and let your front foot shift forward. Your feet do not have to hit together, or your rear foot does not have to knock the front one forward. But this is sometimes a wise practice for the beginner to earn the concept. The reverse is used for going backward. This is an exceptionally good move increasing the gap between you and your opponent in a retreat and to advance upon the opponent for delivering many of the low-line kicks conducive to knife fighting. Think of cutting across all the numbers of the Combat Clock.

    i) Combat Clock sample – left foot at 10. Right foot on axis. Left foot comes to the axis Right foot moves to 5

d) Lunge footwork: Like a fencer, slightly lift your lead foot and propel yourself forward off your rear foot. Do the reverse for going backward. Lift your rear foot and spring back.

e) Lateral footwork: From the fighting stance, if you choose to go a step to the right, then let your right foot step right. Let your left foot follow and move back into a stance. If you choose to go to the left, and then let your left foot step first to the left, then let your right foot follow. Then return to the stance. Try not to cross your feet, for this is a point of imbalance.

> *i.* Combat Clock sample – left foot on axis. Right foot on 3. Right foot moves to axis. Left foot moves to 9.

f) Rocker shuffle: From a fighting stance, slightly bounce your weight on your lead and rear feet shuffling back and forth, or side-to-side. Then change leads in motion and do so from there. Don't over exaggerate the bounce. Stay close to the ground.

g) Sprint forward, back and sides with explosive power. For refined work, sprint out of the Combat Clock using all the clock number.

h) Back peddling: It is important to be able to back quickly away from a situation. In the gym, learn how not to trip over your own two feet. In the real world you must be careful not to trip over objects, but a few clear feet to back peddle can save your life.

9: *The obstacle course:* All of the above mean little without ability to negotiate a challenging set of real world obstacles and terrain. That's why the elite police and military require their exponents to regularly run and traverse obstacle courses. Many athletic endeavors require similar training. Experiment with this by holding a knife through the course.

10: *Now run!* Warriors run. They jog; they dash; they hop; they leap; they zig zag; they move through space. Never stop running as long as your legs still work. Run in all kinds of weather. A warrior toughens his or her soul by experiencing discomfort that comes from running. The residual benefits are also vital. Run while holding a training knife, or any tool you will carry.

Remember to use common sports as hyper-links to fighting skills. At times, fighting footwork resembles rugby, basketball, football and soccer. Remind the soldier that he or she already has training in fighting footwork. through the activities they pursue. Make the connection.

11: *Do not practice barefoot.* Training barefoot is like practicing to ice skate without skates. Wear what you will really wear when you think you may be in combat or attacked. Barefoot training also leads to more injuries to the ankles, toes and feet. Toughening the feet is an abstract, ancient concept.

12: Lastly, strike and kick while running. Combat is walking and running in explosive seconds while fighting. The soldier should practice all fighting techniques from a run or jog, not always from a static, stance position.

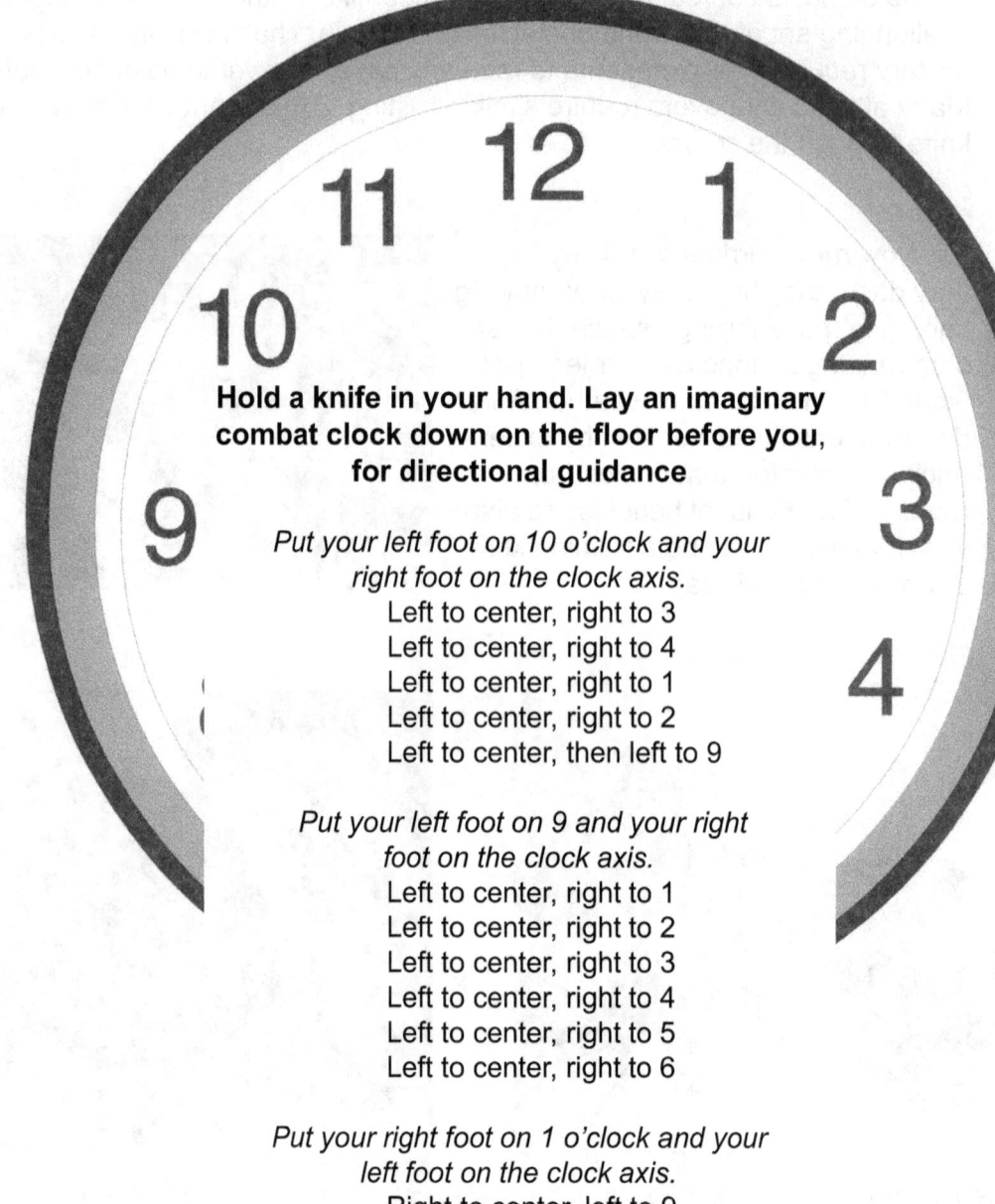

Hold a knife in your hand. Lay an imaginary combat clock down on the floor before you, for directional guidance

Put your left foot on 10 o'clock and your right foot on the clock axis.
Left to center, right to 3
Left to center, right to 4
Left to center, right to 1
Left to center, right to 2
Left to center, then left to 9

Put your left foot on 9 and your right foot on the clock axis.
Left to center, right to 1
Left to center, right to 2
Left to center, right to 3
Left to center, right to 4
Left to center, right to 5
Left to center, right to 6

Put your right foot on 1 o'clock and your left foot on the clock axis.
Right to center, left to 9
Right to center, left to 8
Right to center, left to 10
Right to center, left to 7
Right to center, then right to 3

Put your right foot on 3 and your left foot on the clock axis.
Right to center, left to 11
Right to center, left to 10
Right to center, left to 9
Right to center, left to 8
Right to center, left to 7
Right to center, left to 6

Laid on the ground, the Combat Clock can be utilized for exercising all directions of crawling.

> Crawl Exercise 1: The soldier lays flat on chest, and on command of a number, posts/jumps up and bolts off the clock in the direction of that number.

> Crawl Exercise 2: The soldier lays flat in the center. On command, he crawls off the clock in the direction of the clock number.

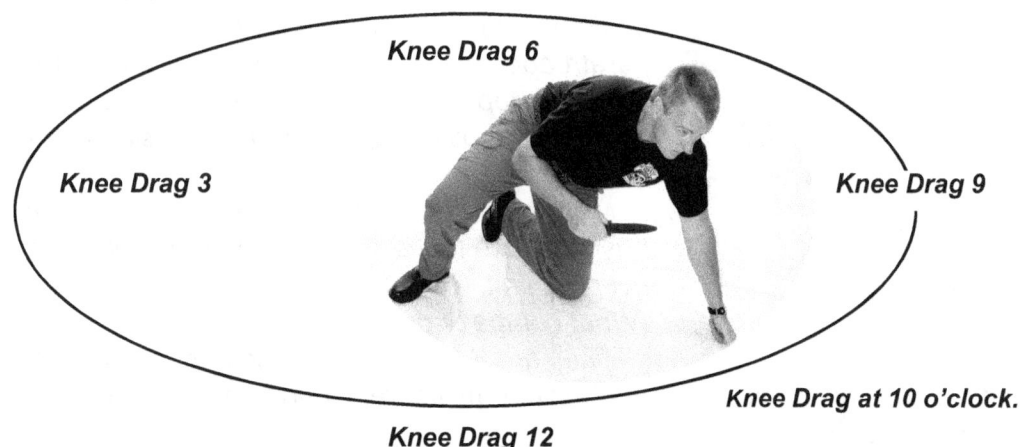

Knee Drag at 10 o'clock.

Laid on the ground, the Combat Clock can be utilized for exercising all directions of knee drags with a knife. A knee drag is an old military exercise that soldiers use when closing in on an enemy. The fist, or elbow/forearm post moves, then the posted knee follows when traveling forward on the clock. Sideward or backward, the knee moves first and the arm follows. Silence is practiced.

> Knee Drag Exercise 1: The soldier "drags" to a called out number on the clock.

To the left, soldiers in the Chinese military practice the knee drag, en masse. here, they are posting with their elbow and using support from their forearm. The knee drag is practiced at various elevations, depending upon the proximity of the enemy.
- Knee/thigh and elbow/forearm (see left).
- Knee and fist/hand (see prior page.

The more body parts touching the ground, the more noise is possible. The term "drag" is a bit of a misnomer. "Lift" would be better to move quietly.

Black Box Knife Combat Files

Unsuspecting Nazi and American units landed in the dark in the same site....

"The newcomers were as startled to detect the dim outlines of the soldiers occupying the riverbed as the paratroopers were to suddenly have newcomers leap nearly on top of them. But none in either group showed any concern over finding themselves side-by-side in the black ditch.

Moments later, a curious newcomer called out in a hushed voice to the men who had occupied the dry riverbed first, "Kompanie zu welchernheist gehosen Sie?" (What company do you belong to?)"

"They are Krauts!" An American shouted as he whipped out his knife and tackled the nearest enemy figure, plunging the sharp point of the weapon into the German.

As if the shout had been a starter's gun at a track meet, the American paratroopers leaped to their feet and charged into the nearest Germans. In seconds, the dry riverbed was a whirling mass of thrashing bodies as opposing patrols fought each other in the blackness with daggers, rifle butts and swinging fists. Loud curses in two languages punctured the night air amid the rustle of entangled bodies rolling in the dust. An occasional grunt could be heard when a trench knife found its mark."

– William B. Breurer,
Drop Zone, Sicily

FORCE NECESSARY: KNIFE!
KNIFE FIGHTING
Knife vs. Hand - Knife vs. Stick - Knife vs. Knife - Knife vs. Gun

The Basics of Knife Attack and Defense
- The Saber Thrust Stab
- The Saber Hook Stab
- The Saber Slash
- The Saber Hack
- The Saber Block
- The Saber Pommel Strike

- The Reverse Grip Thrust Stab
- The Reverse Grip Hook Stab
- The Reverse Grip Slash
- The Saber Block
- The Reverse Grip Pommel Strike

The Support Hand Strikes, Grabs, Blocks and Kicks

Black Box Knife Combat Files

"From the outset, Anders Lassen was clearly eager to use his fighting knife (he was no stranger to knives; during basic training he had demonstrated that by stalking and killing a stag with one, much to the delight of his comrades, who welcomed the fresh meat) and had to be restrained from tackling a sentry quite unnecessarily. Said Les Wright: 'This was my only raid with Andy Lassen, and Apple had already stopped him from doing in a sentry who passed us as we lay down for a few minutes after climbing the Hog's Back.'

Now, as the party approached the Dixcart Hotel, where the twenty-strong German garrison on the island was billeted, Lassen took it upon himself to dispose of the sentinel there. Bombardier Redbom, another member of the party, recalled: 'Andy said he could manage the one sentry on his own. We listened to the German's footsteps and calculated how long it would take him to go back and forth. Andy crept forward alone. The sinister silence was broken by a muffled groan. We looked at each other and guessed what had happened. Then Andy came back and we knew it was alright.'

The German engineers, sent to the island to build a boom defense at Creux harbor, were asleep in single rooms in an annex to the hotel, and within moments of the raiders' entry, were captives.

Wright said: 'We tied their thumbs together and, after cutting their pajama cords, made them hold their trousers up. We were just snatching prisoners for investigation. We didn't sail with handcuffs and manacles.

The prisoners were assembled under cover of trees near the house. In the darkness, one of them suddenly attacked his guard and then, shouting for help and trying to raise the alarm, ran off toward buildings containing a number of Germans. He was caught almost immediately but, after a scuffle, again escaped, still shouting, and was shot. Two of the other prisoners broke away, and both were shot immediately. The fourth, although still held, was accidentally shot in an attempt to silence him by striking him over the head with the butt of a revolver. In fact, it appears that the raider in question 'A bloody big bloke – Toomai the Elephant Boy,' we called him, actually struck the man with the barrel of his pistol, not the butt, forgetting that his finger was on the trigger, 'and blew the top of the man's head off'.

Redbom's memory was of Appleyard, 'shouting, "shut the prisoners up!" This began a regular fight. My prisoner had freed his hands; I bowled him over with a rugby tackle, but he got free again. He was much bigger than me; I couldn't manage him, so I had to shoot him.

Wright covered the party's confused retreat with his Bren gun, and they made their way back to the beach where they had landed, scrambled back into their dory and were soon back aboard 'The Little Pisser', MTB344, an eighteen-metre boat and on their way back to Weymouth.

Lassen layer wrote in his diary, 'The hardest and most difficult job I have ever done – used my knife for the first time.'

A force member who had not been on the raid, Ian Warren, was still asleep when they got back. 'Andy woke me,' he said. 'He held his unwiped knife under my nose, and said, "Look, blood!"

– Roger Ford and Tim Ripley's *The Whites of Their Eyes*

Anders Lassen

CHAPTER 13: THE SABER GRIP ATTACK AND DEFEND

The Saber Stab Knife Angles

There are four stab and slash, hand-position deliveries, controlled by the direction of the hand's palm, that serve as the foundation for stabs, slashes and all follow-up strategies. It's all in the wrist.

        1: The Vertical Blade Angle (palm faces the side).
        2: The 45 Degree Angled Blade (palm slightly angled to the ground).
        3: The Horizontal Blade Angle (palm facing downward).
        4: The Upper Cut Blade Angle (palm facing upward).

The Vertical Blade.

The 45 Degree Angle Blade (right or left angle).

The Horizontal Blade (palm down).

The Uppercut Blade (palm up).

The Saber Grip Thrusting Stabs

The defender will stab at or from the Combat Clock angles. This means that the stab may start from any clock number, or be aimed at any Combat Clock number.

Basic Training: The defender stabs in a straight thrust:
- 12 o'clock or high stab.
- 3 o'clock or stab to the right.
- 6 o'clock or low stab.
- 9 o'clock or stab to the left.
    - lunging thrusts (that do not quickly retract).
    - hit and retract thrusts.
    - upright, walk forward and back, side-to-side.
    - kneeling and seated.
    - grounded.
    - right and left-handed.
    - multiple stabs.

Advanced Training: The defender stabs in a thrust
All 12 numbers on the Combat Clock.
- lunging thrusts.
- hit and retract thrusts.
- upright, walk forward and back, side-to-side.
- kneeling and seated.
- grounded.
- right and left-handed.

Solo Command & Mastery: The Saber Thrusting Stab Basic Training – Standing
Deliver a thrusting saber stab from and/or to each of the four combat clock angles.

High. *Right side.* *Low.* *Left side.*

Solo Command & Mastery: The Saber Thrusting Stab Advanced Training
Deliver a thrusting saber stab from and/or to each of the 12 Combat Clock angles.

Solo Command & Mastery: The Saber Grip Thrust Stabs - Knee High

Lets establish the "rules of the 3 knees." You could be right knee up, both knees down or left knee up.

Left knee up.

Right knee up.

Both knees down.

And there's the "rules of the 3 knees hights." Once knee high (or seated), you will fight people over you, equal to you or under you. Fighting up. Fighting equal. Fighting down.

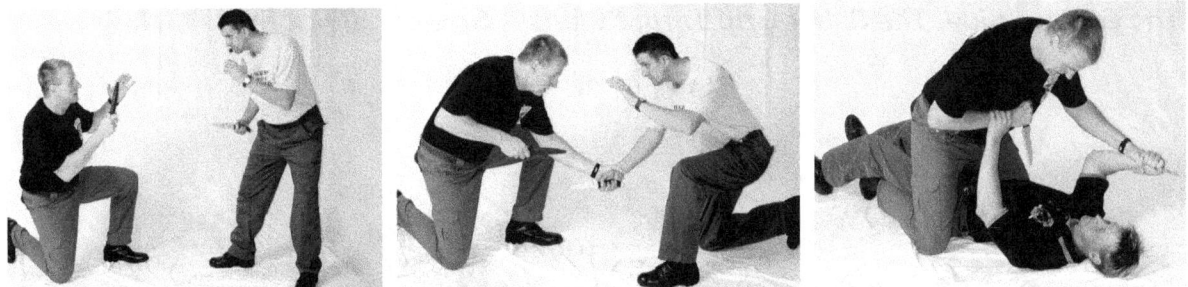
Once knee high, or seated, you will fight people over you, equal to you or under you.

The knee-high defender facing the grounded enemy below him is "top-side," whether he has one knee down or has both knees down. He needs to be aware of these position possibilities when interacting with the enemy. He may or may not use his free support hand to grab or post upon the enemy during this solo practice. Post on your palm, your fist, maybe even your elbow at times.

Both knees down and arm posted.

Both knees down and arms unposted.

With the three heights, a defender can practice the basic and advanced Combat Clock angles to high, equal or low heights. Here is an example of a topside, grounded/floored low basic Combat Clock application.

The Top-Side, The Saber Grip Thrust Stabs – Grounded/Floored

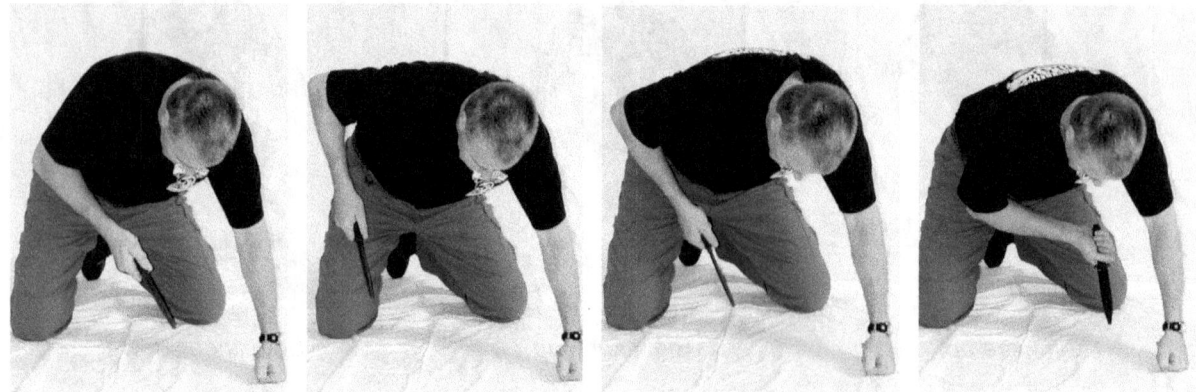

Topside grounded/floored. High (12), right (3), low (6), left (9) solo practice.

The Bottom-Side, The Saber Grip Thrust Stabs – Grounded/Floored

The 12 o'clock or high thrust saber stab.

The 3 o'clock or right thrust saber stab.

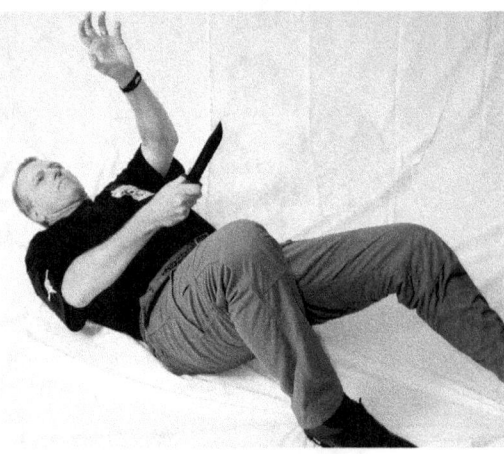

The 6 o'clock or low thrust saber stab.

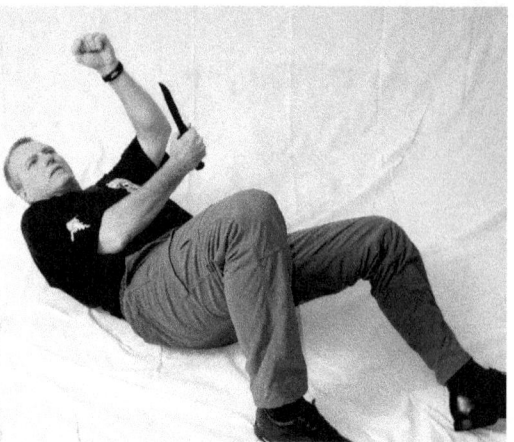

The 9 o'clock or left thrust saber stab.

The Saber Grip Thrust Stabs – Grounded/Floored – On Sides

One of the most important, yet ignored aspects of knife combat is ground fighting, as unrealistic trainers lose themselves in over-emphasizing the duel. When combatants hit the ground from a standing struggle with knives there is a natural propensity to remain on their sides and maintain a non-wrestlers, distance from the edges and tips of the enemy blades. Here are some important solo command and mastery moves from the side. Defenders should practice both right and left sides and should learn the confined, restricted applications of these weapon-side-down-side squeeze tactics.

Weapon-side down! A 12 o'clock high thrusting stab.

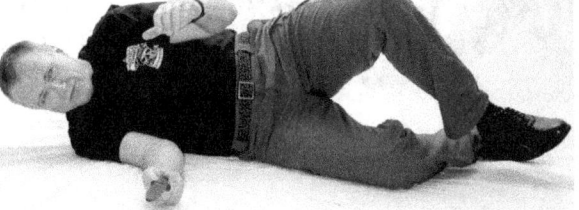
Weapon-side down! A 3 o'clock, right-side thrust stab.

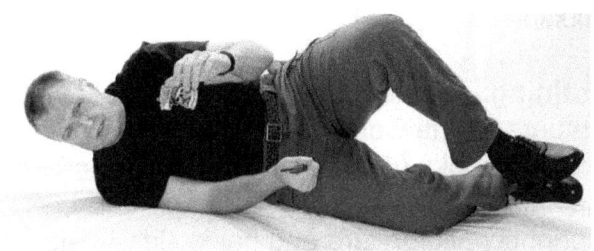
Weapon-side down! A 6 o'clock low thrusting stab.

Weapon-side down! A 9 o'clock, left-side thrust stab.

Weapon-side up! A 12 o'clock high thrusting stab.

Weapon-side up! A 3 o'clock, right-side thrust stab.

Weapon-side up! A 6 o'clock low thrusting stab.

Weapon-side up! A 9 o'clock, right-side thrust stab.

The Saber Grip Hooking Stabs

The defender will stab at or from the Combat Clock angles. This means that the stab may start from any clock number, or be aimed at any Combat Clock number.

Basic Training: The defender stabs in a hook:
- 12 o'clock or high to low hooking stab.
- 3 o'clock or stab, right hooking to the left.
- 6 o'clock or low to high hooking stab.
- 9 o'clock or stab, left hooking to the right.
  - lunging hooks (that do not quickly retract).
  - hit and retract hooks.
  - upright, walk forward and back, side-to-side.
  - kneeling and seated.
  - grounded.
  - right and left-handed.
  - multiples.

Advanced Training: The defender stabs in a hooks:
- All 12 numbers on the Combat Clock.
  - lunging hooks (that do not quickly retract).
  - hit and retract hooks.
  - upright, walk forward and back, side-to-side.
  - kneeling and seated.
  - grounded.
  - right and left-handed.

Solo Command & Mastery: The Saber Hooking Stab Basic Training Standing
Standing. Deliver a thrusting saber stab hooks from and/or to each of the four Combat Clock angles.

12 o' clock or hook from above.

3 o' clock or hook from the right side.

6 o'clock or hook from below.

9 o'clock or hook from the left side.

Solo Command & Mastery: The Saber Hooking Stab, Knee High Sample
Kneeling or seated. Deliver a thrusting saber stab hooks from and/or to each of the four Combat Clock angles.

The 12 o'clock or high saber hook stab.

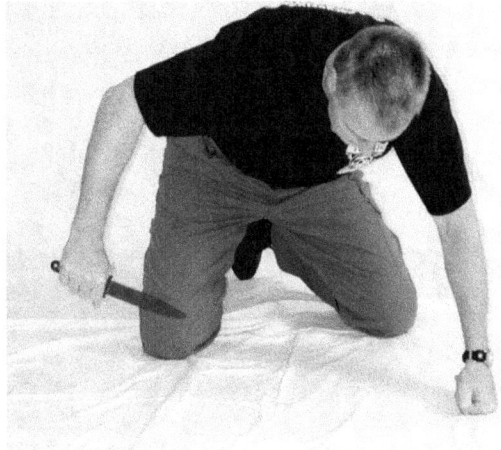
The 3 o'clock or right saber hook stab.

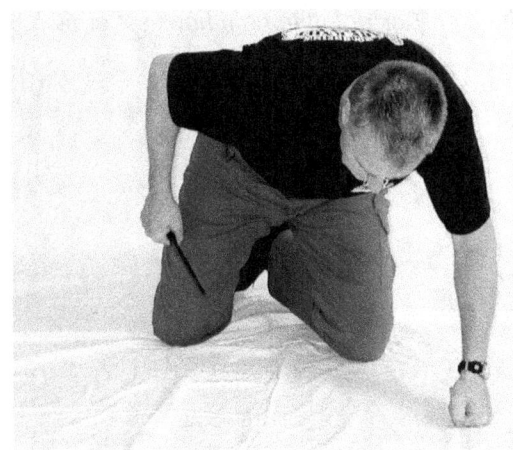
The 6 o'clock or low saber hook stab.

The 9 o'clock or left saber hook stab.

Solo Command & Mastery: The Saber Hooking Stab Basic Training – On Back

Grounded/floored on back. Deliver a thrusting saber stab hooks from and/or to each of the four Combat Clock angles.

The 12 o'clock or high to low hook saber stab.

The 3 o'clock or right to left hook saber stab.

The 6 o'clock or low to high hook saber stab.

The 9 o'clock or left to right hook saber stab.

Solo Command & Mastery: The Saber Grip Hooking Stabs – Grounded/Floored – On Sides

One of the most important, yet ignored aspects of knife combat is ground fighting, as unrealistic trainers lose themselves in over-emphasizing the duel. When combatants hit the ground from a standing struggle with knives there is a natural propensity to remain on their sides and maintain a non-wrestlers, distance from the edges and tips of the enemy blades. Here are some important solo command and mastery moves from the side. Defenders should practice both right and left sides and should learn the confined, restricted applications of these weapon-side-down-side squeeze tactics.

Weapon-side down! A 12 o'clock high hooking stab.

Weapon-side down! A 3 o'clock, right-side thrust stab.

Weapon-side down! A 6 o'clock low hooking stab.

Weapon-side down! A 9 o'clock, left-side thrust stab.

Weapon-side up! A 12 o'clock high hooking stab.

Weapon-side up! A 3 o'clock, right-side hook stab.

Weapon-side up! A 6 o'clock low to high hook stab.

Weapon-side up! A 9 o'clock, left-side hook stab.

Solo Command & Mastery: The Saber Slash

The defender will slash at or from the Combat Clock angles. This means that the stab may start from any clock number, or be aimed at any Combat Clock number.

Basic Training: The defender slashes:
- 12 o'clock or high to low.
- 3 o'clock or from right to left.
- 6 o'clock or low from down to up.
- 9 o'clock or from left to right.
    - lunging thrusts (that do not quickly retract).
    - hit and retract thrusts.
    - upright, walk forward and back, side-to-side.
    - kneeling and seated.
    - grounded.
    - right and left-handed.
    - multiple stabs.
    - combinations

Advanced Training: The defender slashes:
- All 12 numbers on the Combat Clock.
    - lunging thrusts.
    - hit and retract thrusts.
    - upright, walk forward and back, side-to-side.
    - kneeling and seated.
    - grounded.
    - right and left-handed.

The Efficient Slash
The slash is an event at the wrist and elbow, not much at the shoulder. This emphasis makes the slash crisp and efficient, and does not allow overswinging and over-chambering.

The Uncommitted Slash
A slash should have intent, however, if the targeting situation changes, a defender should be able to change course and target "on the fly."

Solo Command & Mastery: The Saber Slash Basic Training – Standing

A slash from 12 downward.   *A slash from the right.*   *A slash up.*   *A slash from the left.*

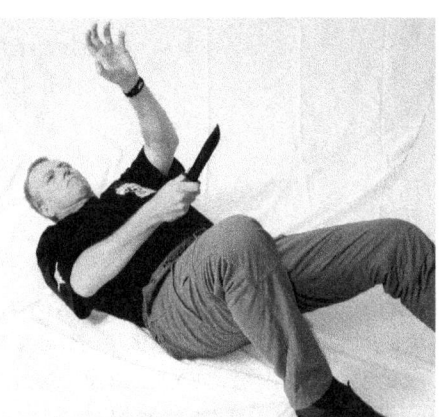

A kneeing slash sample.   *A grounded/floored sample.*

Solo Command & Mastery: The Saber Grip, Chip-out Hack

The defender will saber hack with a twist at or from the Combat Clock angles. This means that the stab may start from any clock number, or be aimed at any Combat Clock number.

Basic Training: The defender hacks in a hook:
12 o'clock or high hack.
3 o'clock or any right side hack.
6 o'clock or low hack.
9 o'clock or left hack.
– hit and retract (a hack is a hit, twist and retract.
– upright, walk forward and back, side-to-side.
– kneeling and seated.
– grounded
– right and left-handed.

Advanced Training: The defender hacks:
All 12 numbers on the Combat Clock.
– upright, walk forward and back, side-to-side.
– kneeling and seated.
– grounded.
– right and left-handed.

Understanding the Hack: The Hacking Chip-Out

The defender can hack and with a sharp and sudden twist of the wrist on impact, create a chipping out motion that can remove a chunk of flesh or bone. It may also be applied when blocking with a knife. An easier task with a saber grip than reverse.

  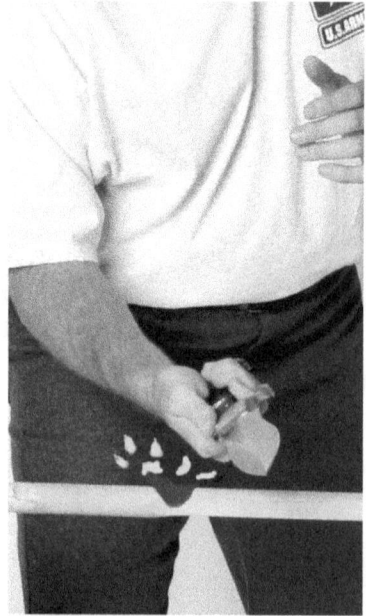

Hit, penetrate and a sharp, quick, twist the wrist for a chip-out.

A powerful hack, twisting chip can cause a impact disarm.

Black Box Knife Combat Files

"All at once Quentin leaped upon the back of the fleeing Boer soldier, as if for a piggy-back ride! And he stuck his dirk into the back and then neck of the absconding man until they tottered and fell. I caught a look of Quentin's face when they rolled, and it was that of a wicked dervish! In all our other frays, Quentin was a model of civilization and of caution. Ten years later at a regimental reunion of sorts, I saw Quentin again, father of two and an assistant shopkeeper. To this very day I still cannot reconcile that brutal afternoon and that savage face I saw, with the now chubby, little man selling garments."

– William Foxtent
  A Volunteer's Diary of the Boer War, Pimpleton

Solo Command & Mastery: Saber Grip Blocking/Hacking

The defender will saber block at or from the Combat Clock angles. This means that the block may start from any clock number, or be aimed at any Combat Clock number. The block position you select will be a result of the position you were in, just before the block is needed. This should be considered a hack if at all possible. A hack is defined as "cut with rough or heavy blows."

Basic Training: The defender blocks/hacks:
12 o'clock or high block, tip points either left or right.
3 o'clock or tight side block, tip up or down.
6 o'clock or low block, tip points right or left.
9 o'clock or left block, tip points up or down,
- upright, walk forward and back, side-to-side.
- kneeling and seated.
- grounded.
- right and left-handed.
- combinations.

Advanced Training: The defender blocks:
All 12 numbers on the Combat Clock.
- upright, walk forward and back, side-to-side.
- kneeling and seated.
- grounded.
- right and left-handed.

Beware the Slashing Block!
The block makes contact, perhaps onto the weapon, perhaps with the weapon-bearing limb? Martial practitioners are often misguided to slash or slash on an incoming attack. This too-fast block/slash does not fully stop the weapon attack. Leave the weapon in place until you are safe in the instant.

The Support Hand Blocks Too!
The Support chapter ahead will cover free hand blocking (and striking.)

High or 12 block.   *Right side or 3 block.*   *Low or 6 block.*   *Left side or 9 block.*

Solo Command & Mastery: The Saber Blocks, Knee High Sample

Kneeling or seated. Deliver a knife block from and/or to each of the four Combat Clock angles, and remember the three knee positions.

Solo Command & Mastery: The Saber Blocks, Knee High, Topside Ground Sample

Kneeling or seated. Deliver a knife block from and/or to each of the four Combat Clock angles.

High or 12 block.   *Right side or 3 block.*   *Low or 6 block.*   *Left side or 9 block.*

Solo Command & Mastery: The Saber Blocks, Bottom-side Ground Sample

Kneeling or seated. Deliver a knife block from and/or to each of the four Combat Clock angles.

  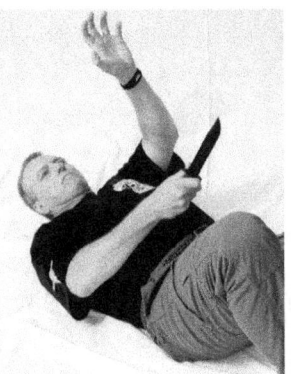 

High or 12 block.   *Right side or 3 block.*   *Low or 6 block.*   *Left side or 9 block.*

Solo Command and Mastery: The Saber Pommel Strike, Thrust or Hook

The defender will pommel strike at or from the Combat Clock angles. This means that the strike may start from any clock number, or be aimed at any Combat Clock number. This section will cover both hooking and thrusting.

Basic Training: The defender pommel strikes:
- 12 o'clock or high.
- 3 o'clock or right side.
- 6 o'clock or low.
- 9 o'clock or left.
    - upright, walk forward and back, side-to-side.
    - kneeling and seated.
    - grounded.
    - right and left-handed.
    - combinations.

Advanced Training: The defender pommel strikes:
All 12 numbers on the Combat Clock.
- upright, walk forward and back, side-to-side.
- kneeling and seated.
- grounded.
- right and left-handed.

The Pommel Strike Briefing

The pommel strike may be implemented with saber and ice pick grips. The pommel strike is a less-than-lethal impact upon the enemy's body. It might stun or set up a capture or kidnapping of an enemy. It may facilitate a more silent attack. The pommel strike might set up a subsequent lethal attack. The pommel of some military knives have been designed with sharp edges or points which causes greater injury than strikes from the common flat pommels. These have been called "skull crushers," and an assortment of other names. If the pommel is designed this way, the soldier cannot affix his thumb atop it, which limits other aspects of knife fighting tactics.

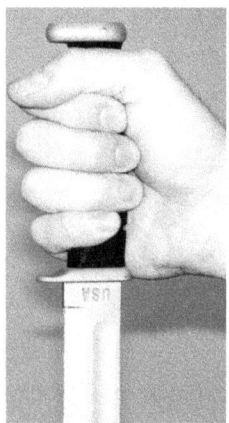 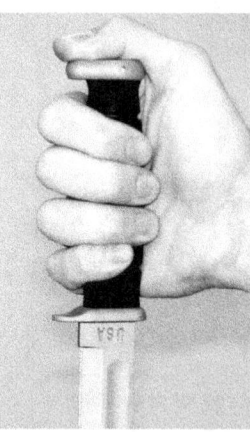 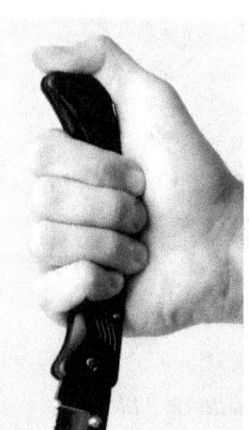  

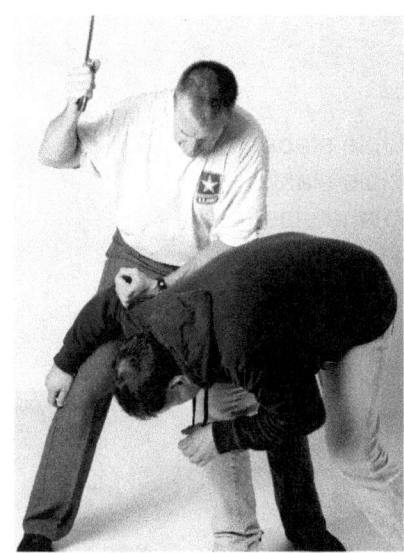

The ability to use the thumb, the palm, even the soldier's body pressing against the pommel, may be an important strategy. Any odd or sharp shaped pommel may impede this tactic.

A properly selected closed folder offers many pommel strike capabilities.

Solo Command and Mastery: Pommel Strike Thrusting and Hooking – Standing
Take care NOT to expose yourself to your own knife tip!

High or 12 pommel.

Right side or 3 pommel.

Low or 6 pommel.

Left side or 9 pommel.

Solo Command & Mastery: Saber Grip Combinations

Saber Combinations 1: Multiple stabs.
Saber Combinations 2: Multiple slashes.
Saber Combinations 3: Stab and slash.
Saber Combinations 4: Slash and Stab.
Saber Combinations 5: Block and slash and stab.
Saber Combinations 6, 7.....invent them.

1: Multiple Stabs: The Chain of Command Drill
This simple drill, exercises the five major saber stab attacks emphasizing the important high, medium and low possibilities. In these 15 quick, drill movements, a defender can learn basic and advanced stabbing and the full potential of the vertical saber stab attack. The progression also follows the common, modern Mixed Martial Arts and Boxing methodologies taught now more all over the world. This really brings out three heights. Stab the links! There are many applications within.

Defender in a Right Lead "Fighting Stance." Pick a lead. Knife in lead side hand.

Series 1:   Jab stab (from lead side).
* high – head.
* medium – torso.
* low – groin.

Series 2:   Cross stab (from rear side).
* high – head.
* medium – torso.
* low – groin.

Series 3:   Hook stab (from lead side.).
* high – head area.
* medium – high torso.
* low – low torso

Series 4:   Uppercut stab (from lead side.).
* high – chin.
* medium – torso.
* low – groin.

Series 5: Descending overhand.
* high – head.
* medium – torso.
* low – low torso.

Repeat with the other hand holding the knife.

Note: *These are important as you will likely be moving in combat. Use this chain in all your grip studies.*

Series 1: The Lead Shoulder (Jab) Stab - high, medium, low. Use these dueling keep-aways.

Series 2: The Rear Shoulder (Cross) Stab - high, medium, low.

Series 3: The Lead Shoulder Hooking Stab - high, medium, low. Use these in a Boxing-style entry,

Series 4: The Uppercut Stab - high, medium, low. Use these in reverse order for a commando drill.

Set 5: The Descending Over Hand - high, medium, low, Use these to get around lead guard arms.

Saber Combinations Series 2: Multiple slashes

Multiple slashes can be served up on the Combat Clock. Here is a famous X Pattern, the Military X Slashes. In the course of military training, instructors often introduce the simple X-pattern of slash/slash/stab. This three-prong attack is frequently touted as "all you need" in a knife fight. Everyone blindly starts from a high right position. While this ultimate proclamation is both ignorant and naive, the slash/slash/stab motion has several practical applications if developed properly.. This X works best with a saber grip as the ice pick slash is gawky and inhibited by right-side high-to-low right and left-side high-to-low, as discussed in the later ice pick slash segment.

The X pattern may be perceived as a three-count motion. Count 1: slash. Count 2: slash. Count 3: stab. Or it may manifest as a one, flowing figure eight pattern with a stab for an end. Instructors may improve the attack skills of their soldiers by showing them both approaches. The flow version helps cure the gawky, static and slow soldier. The 3-count cures the soft, low impact of the over-flowing soldier. The knife is not an orchestra wand. It is an instrument of maiming and death.

Few military and martial instructors cover the X as it should be to maximize the combat applications. Almost all start at high right only! But practice should also start at high left, low right and low left to flesh out the possibilities as this full range of slashes on the X really opens the scope of the Military X tactic. When slashing upward, the soldier must be edge-aware should his knife not be double-edged, turning his hand properly to introduce the edge to the target. Inside the X is the figure moves, and vice versa.

Saber Combinations 3 Grip, Slash and Stab

Sample: A 12 o'clock high slash down and a low stab.

Sample: A 3 o'clock slash from the right and a stab to the left.

Sample: A 6 o'clock slash from below to a high stab.

Sample: A 9 o'clock slash over to the right and a stab.

Saber Combinations 4: Saber Grip, Stab and Slash

Sample: A 12 o'clock high stab. Blade twist. Any slash.

Sample: A 3 o'clock right side stab. Blade twist. Any slash.

Sample: A 6 o'clock low stab. Blade twist. Any slash.

Sample: A 9 o'clock left side stab. Blade twist. Any slash.

In order to do a stab *and* slash, the stab must be free to continue stabbing.

Critical to any stab, is blade evacuation. Knives have been stuck in bodies, especially in bone. A method of removal is the twist. The twist exists within the wrist and at the elbow, which usually works.

Or, the twist is up is the shoulder, which the US Marines once called the "crowbar." I do not know if they still do. This big arm movement, if needed, causes a "chip out" of the substance the knife is stuck in, be it a tree, or...a person.

Wrenching a knife back and forth also can garner a release, but I think a twist is better.

These evacuations cause more grievous injury to the wounds of the enemy.

So, when the defender stabs the enemy, the knife usually travels through clothing and is buried in flesh and bone. At times the blade may become imbedded into the body and to execute a retraction, the defender must manipulate the knife.

Evacuation Manipulation 1: Twisting the handle.
Evacuation Manipulation 2: Pumping the handle inside the wound.
Evacuation Manipulation 3: Combination of both must and probably will be done.
Evacuation Manipulation 4: Power Pull-back, pull-out.

These maneuvers will further open the wound and the path of the knife. Not only will the knife be easier to remove, but it will also create more grievous injury. Forensic specialist will report that once the knife leaves the wound, it creates more opportunity for blood flow.

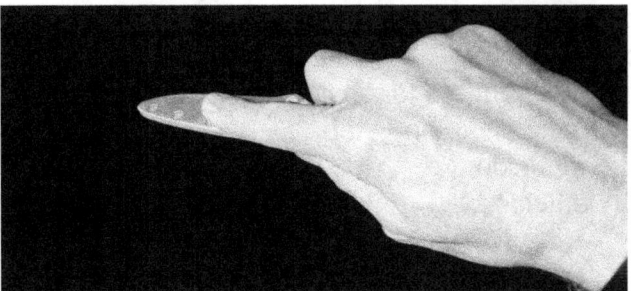

In a proven, forensic world where knife penetration is so very important, how deep do you think this stab will go with this knucklehead's finger high up on the blade like this?

Saber Combinations 5: Block and slash and stab
For your solo training, add a saber knife block/hack within your stabs. Wether it be through the 4 basic or 12 advanced Combat Clock series, add a block before, during or after your sets.

As mentioned earlier, there are support limb blocks that will come in a subsequent chapter.

Your Solo Command & Mastery Saber Grip Workout List

This is a workout list for classroom introduction or refresher. This can be done without gear, or hitting a warpost or heavy bag, or similar gear. Check for athletic synergy. Do right and left hand saber grips. Make sure the support is up and ready for support action.

Saber Thrust, standing.
Saber Thrust, standing, grab and stab.
Saber Thrust, kneeling, low-to-high-standing.
Saber Thrust, kneeling, low to low- kneeling
Saber Thrust, kneeling, topside to ground.
Saber Thrust, seated to high standing.
Saber Thrust, seated to seated.
Saber Thrust, seated to ground.
Saber Thrust, ground, on back, bottom to topside.
Saber Thrust, ground, on back, sit up and stab.
Saber Thrust, ground, on right side.
Saber Thrust, ground, on left side.

Saber Hook Stab, standing.
Saber Hook Stab, standing, grab and stab.
Saber Hook Stab, kneeling, low-to-high-standing.
Saber Hook Stab, kneeling, low to low- kneeling.
Saber Hook Stab, kneeling, topside to ground.
Saber Hook Stab, seated to high standing.
Saber Hook Stab, seated to seated.
Saber Hook Stab, seated to ground.
Saber Hook Stab, ground, on back, bottom to topside.
Saber Hook Stab, ground, on back, sit up and stab.
Saber Hook Stab, ground, on right side.
Saber Hook Stab, ground, on left side.

Saber Slash, standing.
Saber Slash, standing, grab and stab.
Saber Slash, kneeling, low-to-high-standing.
Saber Slash, kneeling, low to low- kneeling
Saber Slash, kneeling, topside to ground.
Saber Slash, seated to high standing.
Saber Slash, seated to seated.
Saber Slash, seated to ground.
Saber Slash, ground, on back, bottom to topside.
Saber Slash, ground, on back, sit up and stab.
Saber Slash, ground, on right side.
Saber Slash, ground, on left side.

Saber Grip Partner Drills

Now we repeat these same solo command and mastery movements with a training partner. The trainer conducts, helps with:
- holding mitts.
- holding shields.
- holds othe helpful gear.
- combat scenarios.
- offers moving targets.
- inspires footwork and mobility.

Slash the stick. Stab the mitt.

There are so many partner drills to work on, standing through the ground, we cannot list them all.

We hope to inspire, not confine. I will cover a few of my favorites that I teach in seminars through the years. But I encourage you review event-based attacks in crime and war and create exercises and drills.

Slash and stab a shield.

Trainer offers a visual pattern for the trainee to begin, here, the X movements.

Sample 1: Face Stab Exercise Skill Drill

Extensive practice and exercise proves that the sudden thrusting stab to the face from a lower target point is extremely difficult to block, stop or deflect. This simple movement is a major, survival, knife fighting tactic. The following is a drill that will increase a soldier's speed, skill, confidence and target acquisition.

Obtain a solid helmet with a solid face shield. Practitioners may have this training gear handy, or may use a motorcycle helmet or a sports helmet. Next, obtain a very soft, training knife.

1) The defender prepares to stab from a ready position. The enemy has his hands up and ready to clap or slap at the stabbing hand. (This is a vicious self-defense stab also, if the practitioner starts from a hands down, common street stance. Concealed knife. He suddenly lunges at the face in this same sudden and direct manner. Practice this version starting with hands down.)

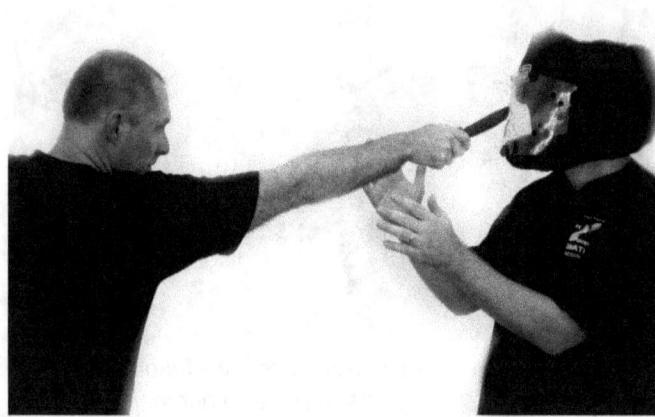

2) The defender thrusts the knife tip at the face shield of the helmet. Limit any telegraphic clues that the attack is coming. The soldier must learn not to tip off the enemy with unnecessary movements like facial expressions, shoulder or hip motions, or grunts or any sounds. If done properly the enemy has trouble stopping the attack.

3) The defender returns the knife to the original position. Exercise this both right and left handed. Continued practice will improve the defender's thrust. Limit any telegraphic clues that the attack is coming.

4) After a series, the soldier steps back and tries from a greater distance. Keep increasing the distance.

5) Do this from ground positions.

6) Improvise new drills based on this concept.

Sample 2: Knife, Stick and Unarmed Strike, Knife Block/Hack
Of course the interaction of strike and block is important. To cover the matrix, the trainer will attack with a knife, a stick, and punches and kicks.

Trainer thrust strikes with knife at 12 high. Trainee blocks/hacks.
Trainer thrust strikes with knife at 3 right. Trainee blocks/hacks.
Trainer thrust strikes with knife at 6 low. Trainee blocks/hacks.
Trainer thrusts strikes with knife at 9 left. Trainee blocks/hacks.
Trainer with knife mixes up the Combat Clock.

Trainer hook strikes with knife at 12 high. Trainee blocks/hacks.
Trainer hook strikes with knife at 3 right. Trainee blocks/hacks.
Trainer hook strikes with knife at 6 low. Trainee blocks/hacks.
Trainer hook strikes with knife at 9 left. Trainee blocks/hacks.
Trainer with knife mixes up the Combat Clock.

Trainer thrust strikes with stick at 12 high. Trainee blocks/hacks.
Trainer thrust strikes with stick at 3 right. Trainee blocks/hacks.
Trainer thrust strikes with stick at 6 low. Trainee blocks/hacks.
Trainer thrusts strikes with stick at 9 left. Trainee blocks/hacks.
Trainer with stick mixes up the Combat Clock.

Trainer hook strikes with stick at 12 high. Trainee blocks/hacks.
Trainer hook strikes with stick at 3 right. Trainee blocks/hacks.
Trainer hook strikes with stick at 6 low. Trainee blocks/hacks.
Trainer hook strikes with stick at 9 left. Trainee blocks/hacks.
Trainer with stick mixes up the Combat Clock.

Trainer attacks with strikes and kicks. Trainee blocks/hacks.
(This is an important self defense drill we call "Versus the Mugger. in which you wound incoming arm and leg attacks and not seriously wound the attacker.)

If the block can also be a hack when possible, this contact might lead to an impact weapon disarm.

Sample 3: The 3-Elevation Drill Strike and Block/Hack Exercise
This is a very comprehensive blocking/hacking exercise. We use this concept through all the *Force Necessary* programs. It is a very quick way to cover high, medium, and low exercises in a great many topics, and by that I mean standing, kneeling, seated and grounded/floored. This develops knife blocking. Start out with any of the mixed weapon groupings listed on the last page. Strike and block.

The signal to change the trainee's height, the moment the trainee knows to drop down a level is by a simulated, leg kick by the trainer. Now, in martial trainee, in systems and arts like MMA, Filipino and Silat to name drop three, great effort is made to detect and avoid being kicked, especially when busy upstairs (standing) with hitting with and without weapons. I always remind practitioners to keep up this kick awareness/avoidance skill, but ..."in this drill"... it's a signal for height-changing.

The exercise starts standing and the trainer and trainee engage in any set of attack or defend. Any subject. When the trainer wants to change heights, the trainer simulates a leg kick, say to the knee. The trainee drops knee high and the same of engagements continue. When the trainer wants to change heights, the trainer simulates another kick to a knee (if one is up) or to the torso. The trainee drops to the ground and the assigned set continues. Next the trainee instigates getting up...SAFELY! The trainee may swing a stick or knife at the trainer, and get up under pressure. Usually we tell the trainee not to turn the back on the trainer.

1) Start out standing. When the trainer wants to change heights, the trainer simulates a leg kick, say to the knee.

2) The trainee drops knee high and the same of engagements continue. When the trainer wants to change heights, the trainer simulates another kick to a knee (if one is up) or to the torso.

3) The trainee drops to the ground and the assigned set continues.

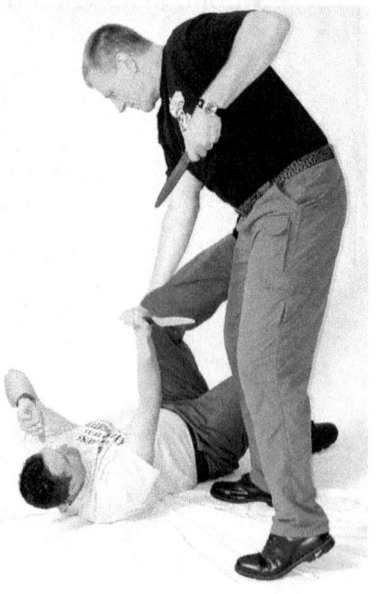

4) Next the trainee instigates getting up...SAFELY! The trainee may swing a stick or knife at the trainer, and get up under pressure. Usually we tell the trainee not to turn the back on the trainer.

Note: When the trainee is grounded, we introduce the option of blocking the knife with a good pair of shoes in a kicking motion.

Note: The trainer may also drop in height along with the trainee, letting the trainee exercise though that experience too.

Your Partner Saber Grip Workout List

Now the trainer holds gear for you to strike! Do right and left hand saber grips. Make sure the support is up and ready for support action.

The stick is for slashing.
The mitt is for stabbing.

    Saber Thrust, standing.
    Saber Thrust, standing, grab and stab.
    Saber Thrust, kneeling, low-to-high-standing.
    Saber Thrust, kneeling, low to low- kneeling
    Saber Thrust, kneeling, topside to ground.
    Saber Thrust, seated to high standing.
    Saber Thrust, seated to seated.
    Saber Thrust, seated to ground.
    Saber Thrust, ground, on back, bottom to topside.
    Saber Thrust, ground, on back, sit up and stab.
    Saber Thrust, ground, on right side.
    Saber Thrust, ground, on left side.

    Saber Hook Stab, standing.
    Saber Hook Stab, standing, grab and stab.
    Saber Hook Stab, kneeling, low-to-high-standing.
    Saber Hook Stab, kneeling, low to low- kneeling
    Saber Hook Stab, kneeling, topside to ground.
    Saber Hook Stab, seated to high standing.
    Saber Hook Stab, seated to seated.
    Saber Hook Stab, seated to ground.
    Saber Hook Stab, ground, on back, bottom to topside.
    Saber Hook Stab, ground, on back, sit up and stab.
    Saber Hook Stab, ground, on right side.
    Saber Hook Stab, ground, on left side.

    Saber Slash, standing.
    Saber Slash, standing, grab and stab.
    Saber Slash, kneeling, low-to-high-standing.
    Saber Slash, kneeling, low to low- kneeling
    Saber Slash, kneeling, topside to ground.
    Saber Slash, seated to high standing.
    Saber Slash, seated to seated.
    Saber Slash, seated to ground.
    Saber Slash, ground, on back, bottom to topside.
    Saber Slash, ground, on back, sit up and stab.
    Saber Slash, ground, on right side.
    Saber Sash, ground, on left side.

Saber Slash and Stab, standing.
Saber Slash and Stab, standing, grab and stab.
Saber Slash and Stab, kneeling, low-to-high-standing.
Saber Slash and Stab, kneeling, low to low- kneeling
Saber Slash and Stab, kneeling, topside to ground.
Saber Slash and Stab, seated to high standing.
Saber Slash and Stab, seated to seated.
Saber Slash and Stab, seated to ground.
Saber Slash and Stab, ground, on back, bottom to topside.
Saber Slash and Stab, ground, on back, sit up and stab.
Saber Slash and Stab, ground, on right side.
Saber Slash and Stab, ground, on left side.

Saber Stab and Slash, standing.
Saber Stab and Slash, standing, grab and stab
Saber Stab and Slash, kneeling, low-to-high-standing.
Saber Stab and Slash, kneeling, low to low- kneeling
Saber Stab and Slash, kneeling, topside to ground.
Saber Stab and Slash, seated to high standing.
Saber Stab and Slash, seated to seated.
Saber Stab and Slash, seated to ground.
Saber Stab and Slash, ground, on back, bottom to topside.
Saber Stab and Slash, ground, on back, sit up and stab.
Saber Stab and Slash, ground, on right side.
Saber Stab and Slash, ground, on left side.

Saber Block/Hack, standing.
Saber Block/Hack, standing, grab and stab.
Saber Block/Hack, kneeling, low-to-high-standing.
Saber Block/Hack, kneeling, low to low- kneeling
Saber Block/Hack, kneeling, topside to ground.
Saber Block/Hack, seated to high standing.
Saber Block/Hack, seated to seated.
Saber Block/Hack, seated to ground.
Saber Block/Hack, ground, on back, bottom to topside.
Saber Block/Hack, ground, on back, sit up and stab.
Saber Block/Hack, ground, on right side.
Saber Block/Hack, ground, on left side.

Chapter 14: The Reverse Grip Attack and Defend

The Reverse Grip Thrusting Stab

The defender will stab at or from the Combat Clock angles. This means that the stab may start from any clock number, or be aimed at any Combat Clock number.

Basic Training: The defender stabs in a straight thrust:
  12 o'clock or high stab.
  3 o'clock or stab to the right.
  6 o'clock or low stab.
  9 o'clock or stab to the left.
   – lunging thrusts (that do not quickly retract).
   – hit and retract thrusts.
   – upright, walk forward and back, side-to-side.
   – kneeling and seated.
   – grounded.
   – right and left-handed.
   – multiple stabs.

Advanced Training: The defender stabs in a thrust
  All 12 numbers on the Combat Clock.
   – lunging thrusts.
   – hit and retract thrusts.
   – upright, walk forward and back, side-to-side.
   – kneeling and seated.
   – grounded.
   – right and left-handed.

Solo Command & Mastery: The Reverse Grip Thrusting Stab Basic Training – Standing

Deliver a thrusting saber stab from and/or to each of the four combat clock angles.

12 or high.   *3 or from the right.*   *6 or low.*   *9 or from the left.*

Solo Command & Mastery: The Reverse Grip Thrust Stabs – Knee High

Lets review the "rules of the 3 knees." You could be right knee up, both knees down or left knee up.

And we review the "rules of the 3 knees hights." Once knee high (or seated), you will fight people over you, equal to you or under you. Fighting up. Fighting equal. Fighting down.

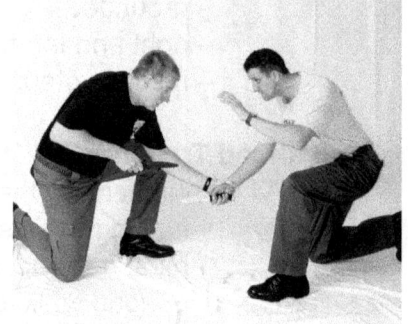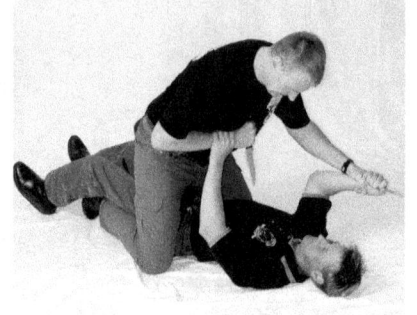

The knee-high defender facing the grounded enemy below him is "top-side," whether he has one knee down or has both knees down. He needs to be aware of these position possibilities when interacting with the enemy. He may or may not use his free support hand to grab or post upon the enemy during this solo practice. Post on your palm, your fist, maybe even your elbow at times. With the three heights, a defender can practice the basic and advanced Combat Clock angles to high, equal or low heights.

The Top-Side, The Reverse Grip Thrust Stabs – Grounded/Floored

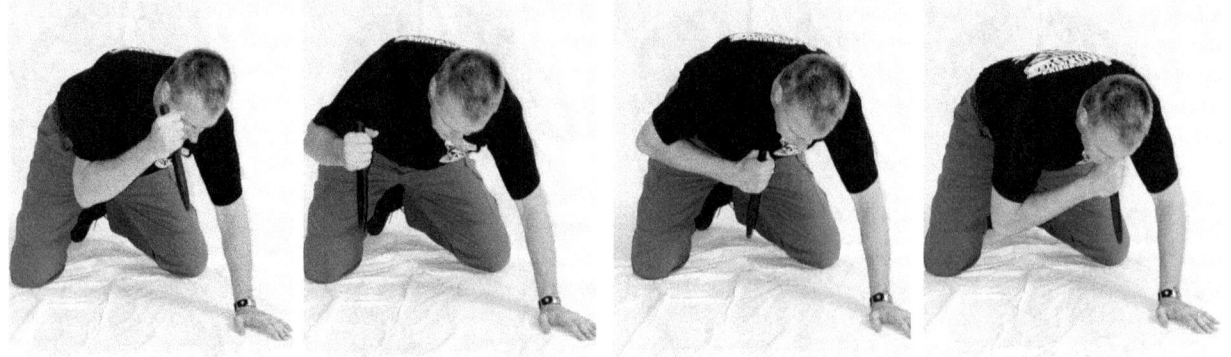

12 or high.   *3 or from the right.*   *6 or low.*   *9 or from the left.*

Solo Command and Mastery Reverse Grip – Bottom-side

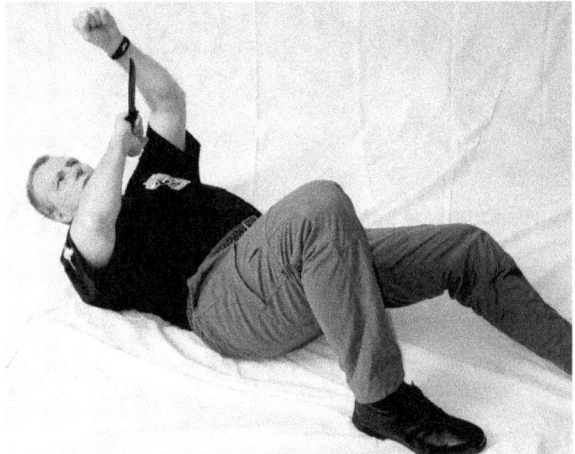
The 12 o'clock or high thrust stab.

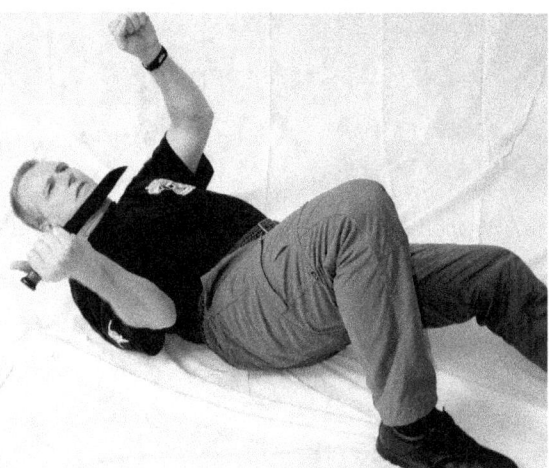
The 3 o'clock or right side pick stab.

The 6 o'clock or low thrust stab.

The 9 o'clock or left side stab.

Solo Command and Mastery Reverse Grip – On Sides

Weapon-side down! A 12 o'clock high thrusting stab.

Weapon-side down! A 3 o'clock, right-side thrust.

Weapon-side down! A 6 o'clock low thrusting stab.

Weapon-side down! A 9 o'clock, left-side thrust stab.

Weapon-side up! A 12 o'clock high thrusting stab.

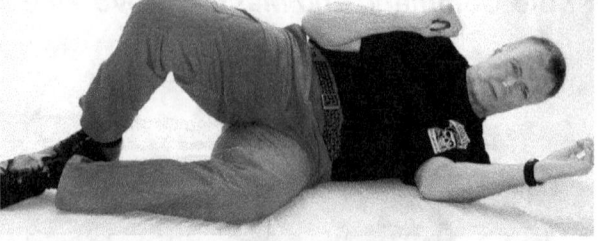
Weapon-side up! A 3 o'clock, right-side thrust stab.

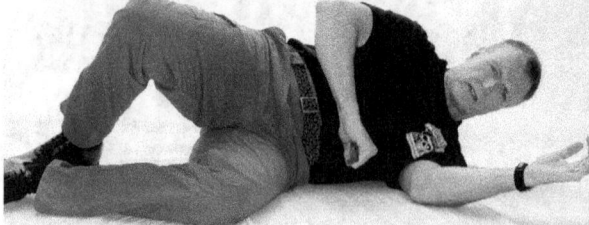
Weapon-side up! A 6 o'clock low thrusting stab.

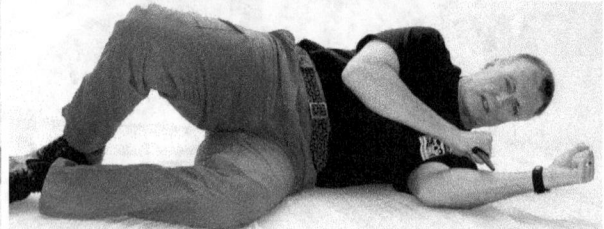
Weapon-side up! A 9 o'clock, right-side thrust stab.

Reverse Grip Note:
Numerous knife systems offer a palm-to-pommel support hand. Power and support!

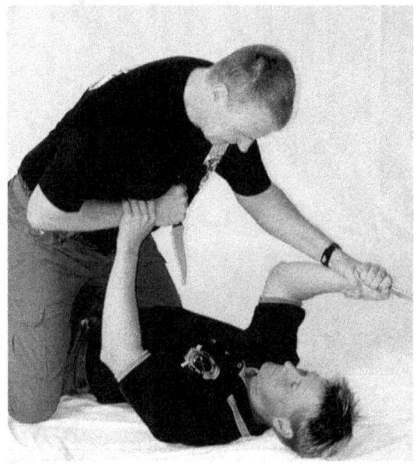

The Reverse Grip Hooking Stabs

The defender will stab at or from the Combat Clock angles. This means that the stab may start from any clock number, or be aimed at any Combat Clock number.

Basic Training: The defender stabs in a hooking:
- 12 o'clock or high stab.
- 3 o'clock or stab to the right.
- 6 o'clock or low stab.
- 9 o'clock or stab to the left.
    - lunging thrusts (that do not quickly retract).
    - hit and retract thrusts.
    - upright, walk forward and back, side-to-side.
    - kneeling and seated.
    - grounded.
    - right and left-handed.
    - multiple stabs.

Advanced Training: The defender stabs in a hook
- All 12 numbers on the Combat Clock.
    - lunging thrusts.
    - hit and retract thrusts.
    - upright, walk forward and back, side-to-side.
    - kneeling and seated.
    - grounded.
    - right and left-handed.

Solo Command & Mastery: The Reverse Grip Hook Stab Basic Training – Standing
Deliver a thrusting saber stab from and/or to each of the four combat clock angles.

12 or high. *3 or from the right.* *6 or low.* *9 or from the left.*

Solo Command & Mastery: The Reverse Grip Hook Stab Basic Training – Kneeling

We fight people over us, equal to us, and below us, as with this example.

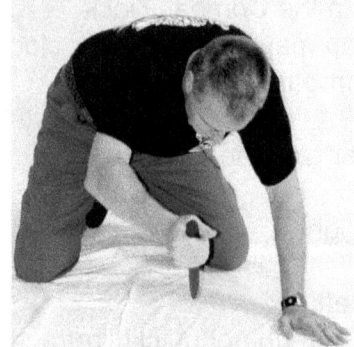
The 12 o'clock or high hook stab.

The 3 o'clock or right side stab.

The 6 o'clock or low hook stab.

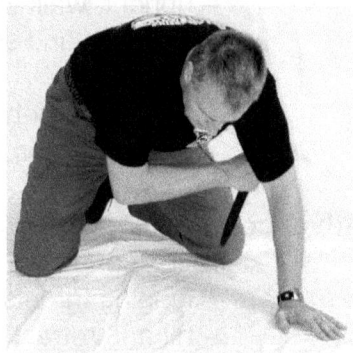
The 9 o'clock or left side hook stab.

Solo Command & Mastery: The Reverse Grip Hook Stab Basic Training – Bottom-side

12 or high.

3 or from the right.

6 or low.

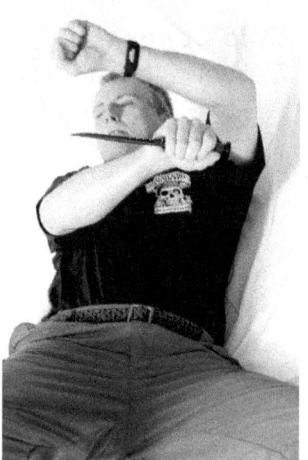
9 or from the left.

Solo Command & Mastery: The Reverse Grip Hook Stab Basic Training – On-side

Weapon-side down! A 12 o'clock high hooking stab.

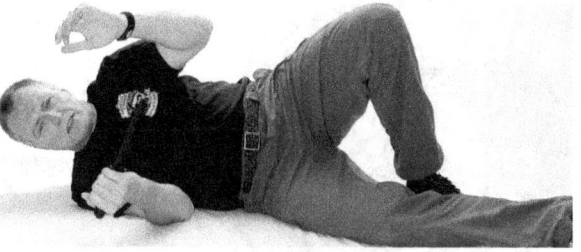

Weapon-side down! A 3 o'clock, right-side hook stab.

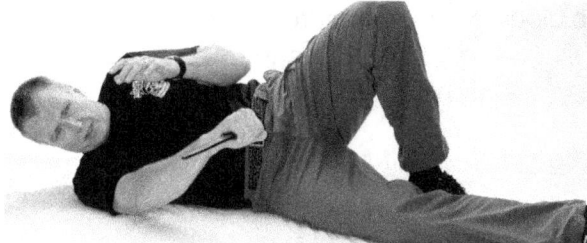

Weapon-side down! A 6 o'clock low hooking stab.

Weapon-side down! A 9 o'clock, left-side hook stab.

Weapon-side up! A 12 o'clock high hooking stab.

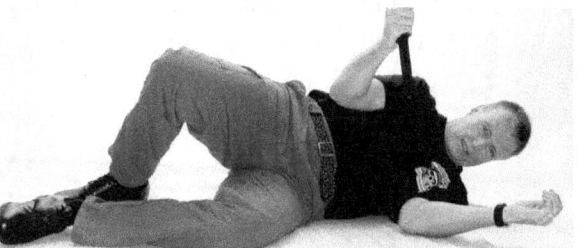

Weapon-side up! A 3 o'clock, right-side hook stab.

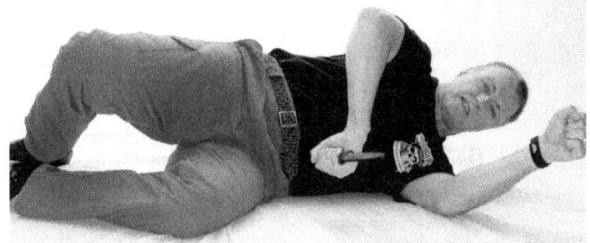

Weapon-side up! A 6 o'clock low hooking stab.

Weapon-side up! A 9 o'clock, right-side hook stab.

The Reverse Grip Slash

The defender will slash at or from the combat clock angles. This means that the slash may start from a clock number, or be aimed at a clock number. Keep the slash fast and concise. No over-chambering.

Basic Training: The defender efficiently slashes-
12 o'clock or any slash from above.
3 o'clock or any slash from the right side.
6 o'clock or any from below
9 o'clock or any slash from the left side.
– right and left handed.
– standing through ground/floor.

Advanced Training:
Ordinarily, the program now calls for slashing on all 12 numbers of the Combat Clock at this point. However, in the case of the reverse knife grip, it become powerless and useless to strike on all 12 angles. The right hand can deliver solid reverse grip slashes from 1 o'clock to about 9 o'clock. On the remaining "15 minutes" or quarter of the clock, the slashes become lame and awkward due to the unusual arm position.

Right-handed Reverse Grip
Difficulty slashing on 10, 11, 12

Left-handed Ice Pick Grip
Difficulty slashing on 12, 1, 2

Conversely, the left hand can deliver solid reverse grip slashes from about 3 o'clock on to about 12 o'clock. On the remaining "15 minutes" or a quarter of the clock, the slashes become lame and awkward due to the unusual arm position. Reverse grip slashes often matches the arm mechanics of an elbow strike.

The Efficient Slash
The slash is an event at the wrist and elbow, not much at the shoulder. This emphasis makes the slash crisp and efficient, and does not allow overswinging and over-chambering.

The Uncommitted Slash

A slash should have intent, however, if the targeting situation changes, a defender should be able to change course and target "on the fly."

Solo Command & Mastery Reverse Grip Slash – Standing

12 or high.   3 or from the right.   6 or low.   9 or from the left.

Solo Command & Mastery: The Reverse Grip Slash Basic Training – Kneeling

We fight people over us, equal to us, and below us, as with these examples.

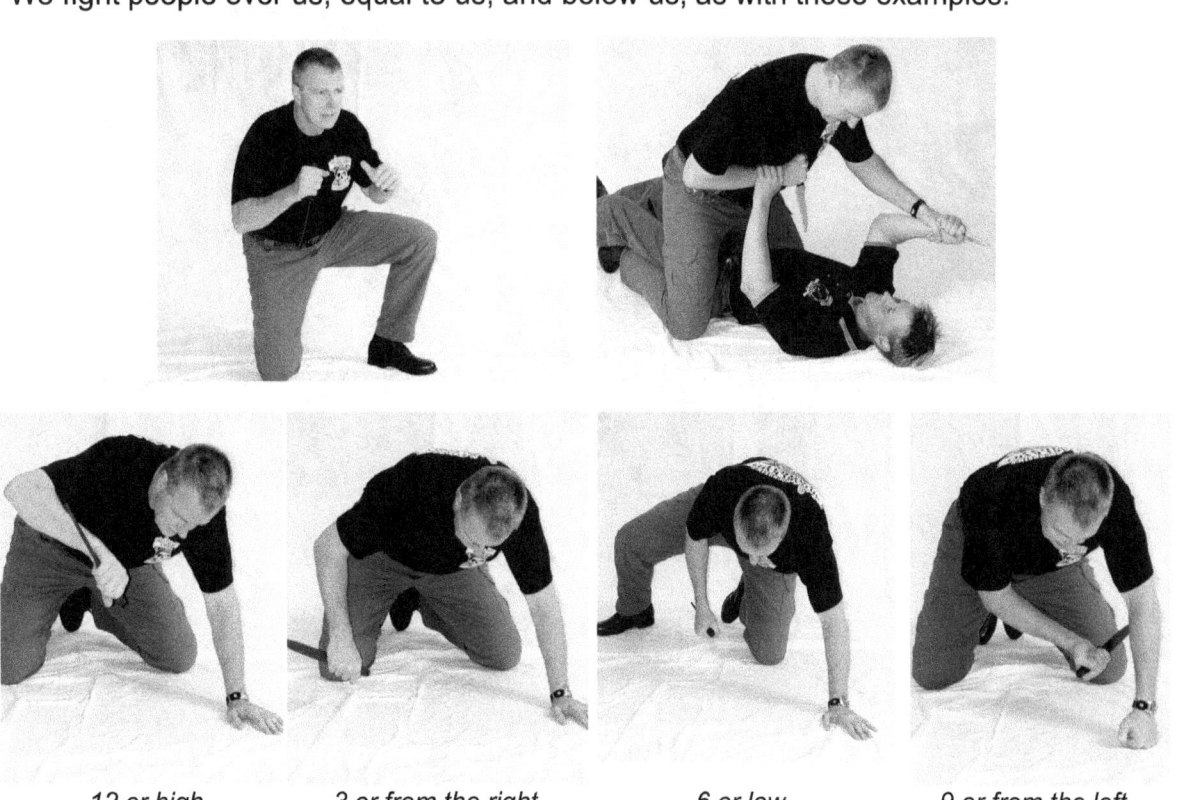

12 or high.   3 or from the right.   6 or low.   9 or from the left.

Solo Command & Mastery: The Reverse Grip Slash Basic Training – Bottom-side

12 o'clock, high to low slash.

The 3 o'clock or right to left slash.

The 6 o'clock or low to high slash.

The 9 o'clock or left to right slash.

Solo Command & Mastery: The Reverse Grip Slash Basic Training – On-sides

Weapon-side down! A 12 o'clock high to low slash.

Weapon-side down! A 3 o'clock, right to left slash.

Weapon-side down! A 6 o'clock low to high slash.

Weapon-side down! A 9 o'clock, left to right slash.

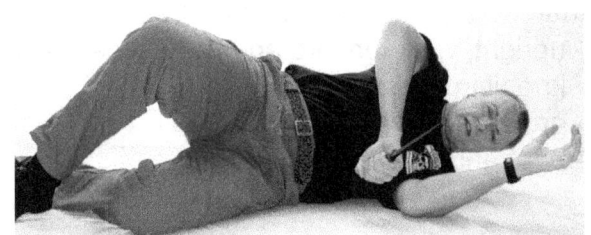
Weapon-side up! A 12 o'clock high to low slash.

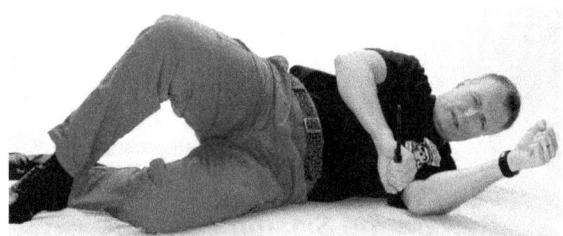
Weapon-side up! A 3 o'clock, right to left slash.

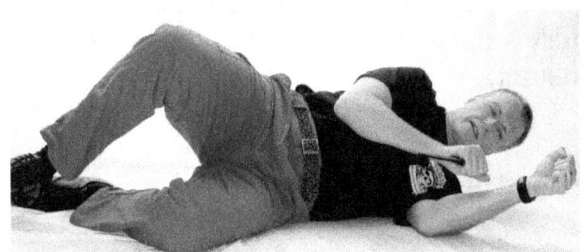
Weapon-side up! A 6 o'clock low to high slash.

Weapon-side up! A 9 o'clock, left to right side slash.

Solo Command & Mastery Blocking/Hacking

The defender will reverse grip block at or from the Combat Clock angles. This means that the block may start from any clock number, or be aimed at any Combat Clock number. The block position you select will be a result of the position you were in, just before the block is needed. This should be considered a hack if at all possible. A hack is defined as "cut with rough or heavy blows."

Basic Training: The defender blocks/hacks:
12 o'clock or high block, tip points either left or right.
3 o'clock or tight side block, tip up or down.
6 o'clock or low block, tip points right or left.
9 o'clock or left block, tip points up or down,
– upright, walk forward and back, side-to-side.
– kneeling and seated.
– grounded.
– right and left-handed.
– combinations.

Advanced Training: The defender blocks:
All 12 numbers on the Combat Clock.
– upright, walk forward and back, side-to-side.
– kneeling and seated.
– grounded.
– right and left-handed.

Beware the Slashing Block!
The block makes contact, perhaps onto the weapon, perhaps with the weapon-bearing limb? Martial practitioners are often misguided to slash or slash on an incoming attack. This too-fast block/slash does not fully stop the weapon attack. Leave the weapon in place until you are safe in the instant.

The Support Hand Blocks Too!
The Support chapter ahead will cover free hand blocking (and striking.)

Solo Command & Mastery: The Reverse Grip Slash Basic Training – Kneeling
We fight people over us, equal to us, and below us, as with these examples.

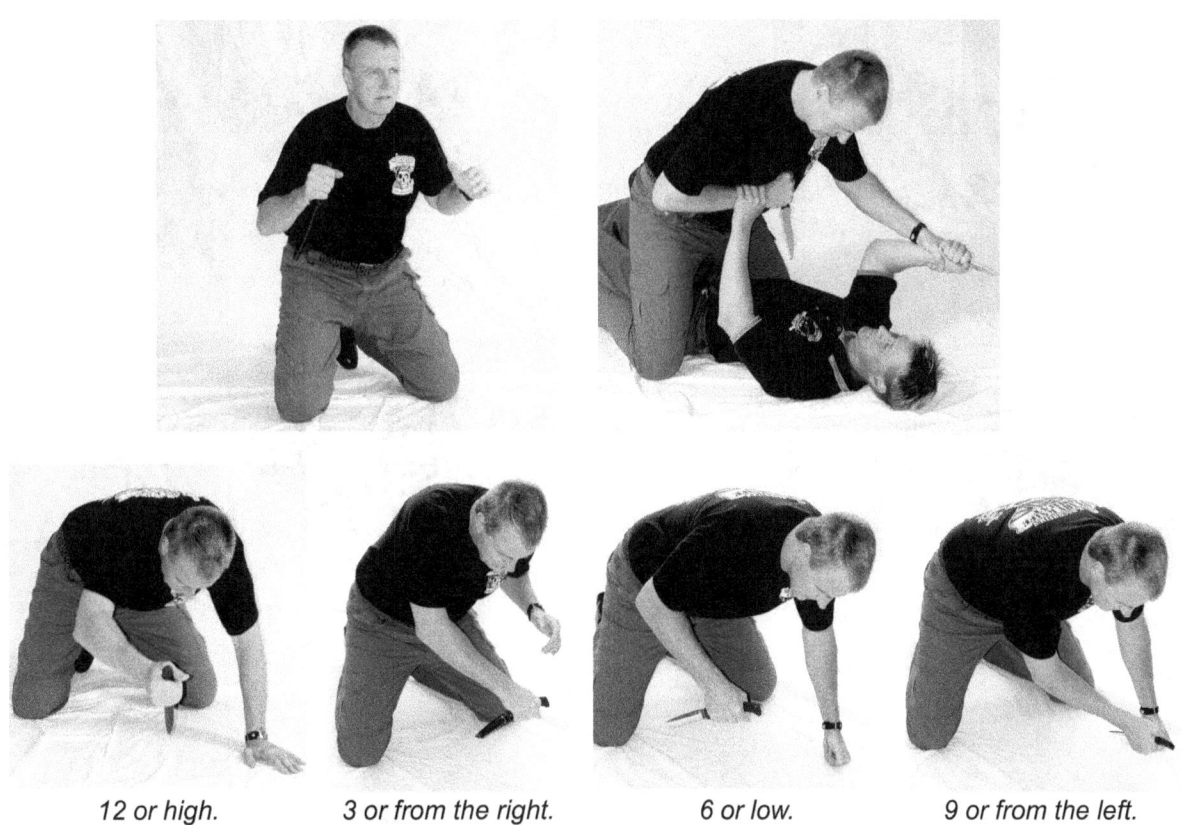

12 or high.   3 or from the right.   6 or low.   9 or from the left.

Solo Command & Mastery: The Reverse Grip Slash Basic Training – Bottom-side

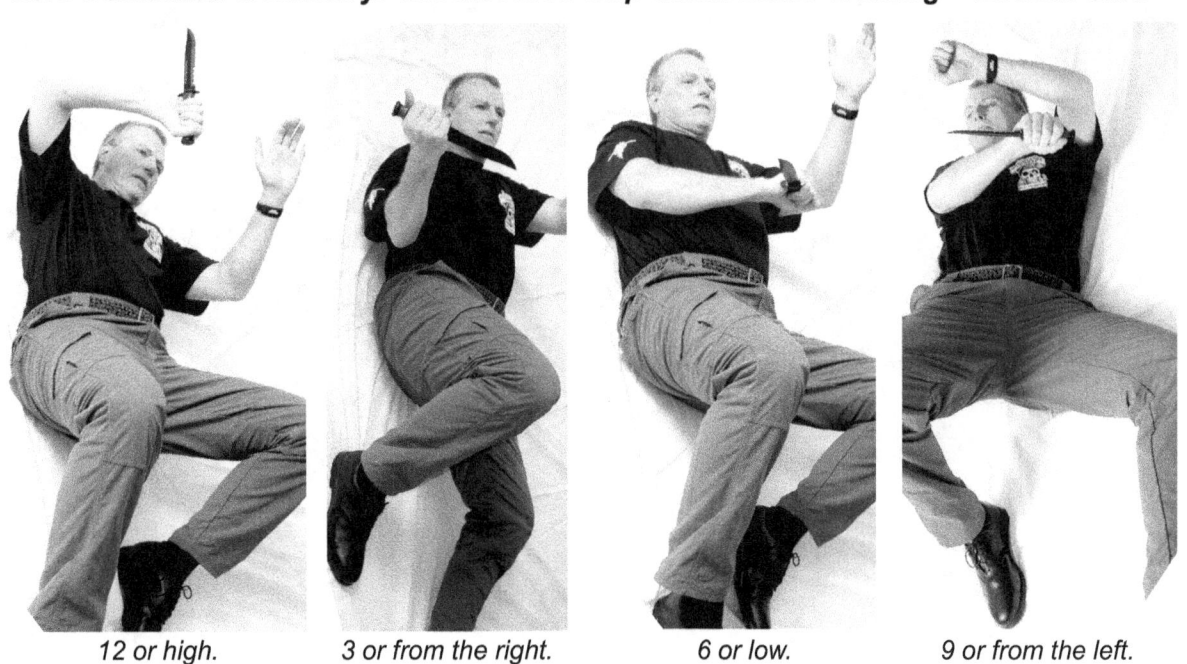

12 or high.   3 or from the right.   6 or low.   9 or from the left.

Command & Mastery Grounded – On Sides

Weapon-side down! A 12 o'clock high hack.

Weapon-side down! A 3 o'clock, right-side hack.

Weapon-side down! A 6 o'clock hack.

Weapon-side down! A 9 o'clock, left-side hack.

Weapon-side up! A 12 o'clock high hack.

Weapon-side up! A 3 o'clock, right-side hack.

Weapon-side up! A 6 o'clock hack.

Weapon-side up! A 9 o'clock hack.

Solo Command & Mastery Reverse Grip Pommel Strikes Thrusts and Hooks

The defender will pommel strike at or from the Combat Clock angles. This means that the strike may start from any clock number, or be aimed at any Combat Clock number. This section will cover both hooking and thrusting.

Basic Training: The defender pommel strikes:
  12 o'clock or high.
  3 o'clock or right side.
  6 o'clock or low.
  9 o'clock or left.
– upright, walk forward and back, side-to-side.
– kneeling and seated.
– grounded.
– right and left-handed.
– combinations.

Advanced Training: The defender pommel strikes:
  All 12 numbers on the Combat Clock.
– upright, walk forward and back, side-to-side.
– kneeling and seated.
– grounded.
– right and left-handed.

12 or high.   3 or from the right.   6 or low.   9 or from the left.

Solo Command & Mastery Reverse Grip Pommel Strikes - Kneeling
We fight people over us, equal to us, and below us, as with these examples.

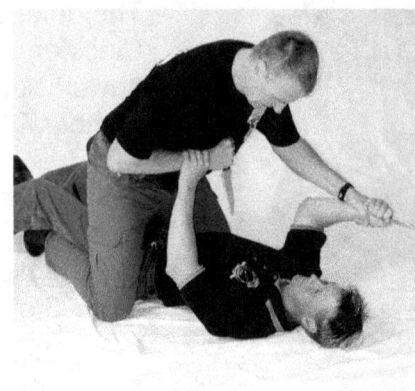

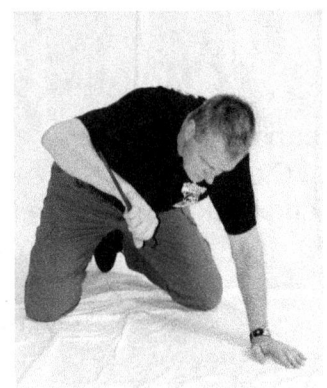

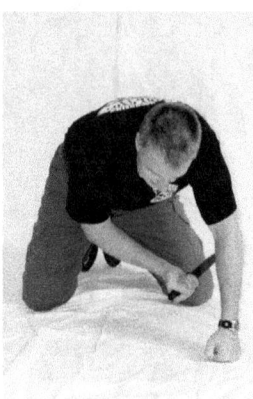

12 or high.   3 or from the right.   6 or low.   9 or from the left.

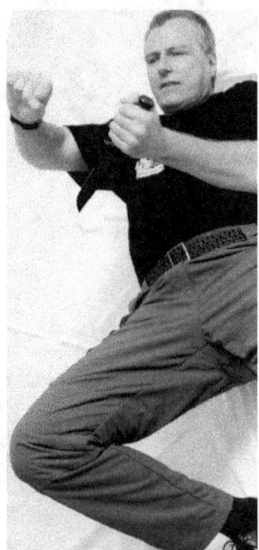

12 or high.   3 or from the right.   6 or low.   9 or from the left.

Solo Command & Mastery: Reverse Grip Combinations

Reverse Grip Combinations 1: Multiple stabs.
Reverse Grip Combinations 2: Multiple slashes.
Reverse Grip Combinations 3: Stab and slash.
Reverse Grip Combinations 4: Slash and Stab.
Reverse Grip Combinations 5: Block-Slash-Stab.
Reverse Grip Combinations 6: Pommel Strike and Slash.
Reverse Grip Combinations Invent them.

Imagine the combinations of slashing and stabbing. Run the numbers of multiple attacks.

Your Solo Command & Mastery Saber Grip Workout List

This is a workout list for classroom introduction or refresher. This can be done without gear, or hitting a warpost or heavy bag, or similar gear. Check for athletic synergy. Do right and left hand saber grips. Make sure the support is up and ready for support action.

    Reverse Grip Thrust, standing.
    Reverse Grip Thrust, standing, grab and stab.
    Reverse Grip Thrust, kneeling, low-to-high-standing.
    Reverse Grip Thrust, kneeling, low to low- kneeling
    Reverse Grip Thrust, kneeling, topside to ground.
    Reverse Grip Thrust, seated to high standing.
    Reverse Grip Thrust, seated to seated.
    Reverse Grip Thrust, seated to ground.
    Reverse Grip Thrust, ground, on back, bottom to topside.
    Reverse Grip Thrust, ground, on back, sit up and stab.
    Reverse Grip Thrust, ground, on right side.
    Reverse Grip Thrust, ground, on left side.

    Reverse Grip Hook Stab, standing.
    Reverse Grip Hook Stab, standing, grab and stab
    Reverse Grip Hook Stab, kneeling, low-to-high-standing.
    Reverse Grip Hook Stab, kneeling, low to low- kneeling
    Reverse Grip Hook Stab, kneeling, topside to ground.
    Reverse Grip Hook Stab, seated to high standing.
    Reverse Grip Hook Stab, seated to seated.
    Reverse Grip Hook Stab, seated to ground.
    Reverse Grip Hook Stab, ground, on back, bottom to topside.
    Reverse Grip Hook Stab, ground, on back, sit up and stab.
    Reverse Grip Hook Stab, ground, on right side.
    Reverse Grip Hook Stab, ground, on left side.

Reverse Grip Slash, standing.
Reverse Grip Slash, standing, grab and stab
Reverse Grip Slash, kneeling, low-to-high-standing.
Reverse Grip Slash, kneeling, low to low- kneeling
Reverse Grip Slash, kneeling, topside to ground.
Reverse Grip Slash, seated to high standing.
Reverse Grip Slash, seated to seated.
Reverse Grip Slash, seated to ground.
Reverse Grip Slash, ground, on back, bottom to topside.
Reverse Grip Slash, ground, on back, sit up and stab.
Reverse Grip Slash, ground, on right side.
Reverse Grip Slash, ground, on left side.

Reverse Grip Pommel Strike and Slash.

Reverse Grip Partner Drills
Now we repeat these same solo command and mastery movements with a training partner. There are so many partner drills to work on, standing through the ground, we cannot list them all. We hope to inspire, not confine.I will cover a few of my favorites that I teach in seminars through the years. But I encourage you review event-based attacks in crime and war and create exercises and drills. Of course more drills and scenarios fill the rest of the book. These are just some beginner's kickstart samples. The trainer conducts, helps with:
- holding mitts.
- holding shields.
- holds othe helpful gear.
- combat scenarios.
- offers moving targets.
- inspires footwork and mobility.

The mitt is for stabbing. The stick is for slashing.

Sample 1: Knife, Stick and Unarmed Strike, Knife Block/Hack
Of course the interaction of strike and block is important. To cover the matrix, the trainer will attack with a knife, a stick, and punches and kicks.

Trainer Thrust Strikes with knife at 12 high.Trainee blocks/hacks.
Trainer Thrust Strikes with knife at 3 right. Trainee blocks/hacks.
Trainer Thrust Strikes with knife at 6 low. Trainee blocks/hacks.
Trainer Thrust Strikes with knife at 9 left. Trainee blocks/hacks.
Trainer with knife mixes up the Combat Clock.

Trainer Hook Strikes with knife at 12 high.Trainee blocks/hacks.
Trainer Hook Strikes with knife at 3 right. Trainee blocks/hacks.
Trainer Hook Strikes with knife at 6 low. Trainee blocks/hacks.

Trainer Hook Strikes with knife at 9 left. Trainee blocks/hacks.
Trainer with knife mixes up the Combat Clock.
Trainer Thrust Strikes with stick at 12 high. Trainee blocks/hacks.
Trainer Thrust Strikes with stick at 3 right. Trainee blocks/hacks.
Trainer Thrust Strikes with stick at 6 low. Trainee blocks/hacks.
Trainer Thrust Strikes with stick at 9 left. Trainee blocks/hacks.
Trainer with stick mixes up the Combat Clock.

Trainer Hook Strikes with stick at 12 high. Trainee blocks/hacks.
Trainer Hook Strikes with stick at 3 right. Trainee blocks/hacks.
Trainer Hook Strikes with stick at 6 low. Trainee blocks/hacks.
Trainer Hook Strikes with stick at 9 left. Trainee blocks/hacks.
Trainer with stick mixes up the Combat Clock.

Trainer attacks with strikes and kicks. Trainee blocks/hacks.
(This is an important self defense drill we call "Versus the Mugger. in which you wound incoming arm and leg attacks and not seriously wound the attacker.)

Sample 2: The 3-Elevation Drill Strike and Block/Hack Exercise
This is a very comprehensive blocking/hacking exercise. We use this concept through all the *Force Necessary* programs. It is a very quick way to cover high, medium, and low exercises in a great many topics, and by that I mean standing, kneeling, seated and grounded/floored. This develops knife blocking. Start out with any of the mixed weapon groupings listed on the last page. Strike and block.

The signal to change the trainee's height, the moment the trainee knows to drop down a level is by a simulated, leg kick by the trainer. Now, in martial trainee, in systems and arts like MMA, Filipino and Silat to name drop three, great effort is made to detect and avoid being kicked, especially when busy upstairs (standing) with hitting with and without weapons. I always remind practitioners to keep up this kick awareness/avoidance skill, but ..."in this drill"... it's a signal for height-changing.

The exercise starts standing and the trainer and trainee engage in any set of attack or defend. Any subject. When the trainer wants to change heights, the trainer simulates a leg kick, say to the knee. The trainee drops knee high and the same of engagements continue. When the trainer wants to change heights, the trainer simulates another kick to a knee (if one is up) or to the torso. The trainee drops to the ground and the assigned set continues. Next the trainee instigates getting up...SAFELY! The trainee may swing their knife back at the trainer, and get up under pressure. Usually we tell the trainee not to turn the back on the trainer.

1: Start out standing. When the trainer wants to change heights, the trainer simulates a leg kick, say to the knee.

2: The trainee drops knee high and the same of engagements continue. When the trainer wants to change heights, the trainer simulates another kick to a knee (if one is up) or to the torso.

3: The trainee drops to the ground and the assigned set continues.

4: Next the trainee instigates getting up...SAFELY! The trainee may swing a stick or knife at the trainer, and get up under pressure. Usually we tell the trainee not to turn their back on the trainer.

Note: When the trainee is grounded, we introduce the option of blocking the knife with a good pair of shoes in a kicking motion.

Note: The trainer may also drop in height along with the trainee, letting the trainee exercise though that experience too.

CHAPTER 15: SUPPORT! HAND STRIKES, GRABS, BLOCKS AND KICKS
Solo Command & Mastery: The Support Limb Blocking

Blocking whilst holding a knife exists in ancient, military training records. Modern training manuals of the 20th and 21st century have many isolated photos and drawings of the knife used to block another knife attack, a stick attack or a bayonet attack. While blocking versus a hand, stick, knife or improvised weapon attack is a necessary combat skill it is a skill barely taught in an official and complete capacity in the modern training programs of most militaries. Most training is spent dueling and knife sparring.

We will cycle the empty-hand/unarmed blocking and knife blocking methods through the Combat Clock corners as an unforgettable way to practice the process. It should be noted that there is a mantra amongst some martial courses, *"We never block! We always strike."* And that is a fine motto, but the realities of rabid combat do not always allow us to express this full aggressive intent versus every single incoming attack, but rather often just lets us barely escape contact with some partial protection move like a simple, reflexive block. These blocks deflect straight line attacks and stop hooking attacks. Leave the block in place until the attack stroke is over.

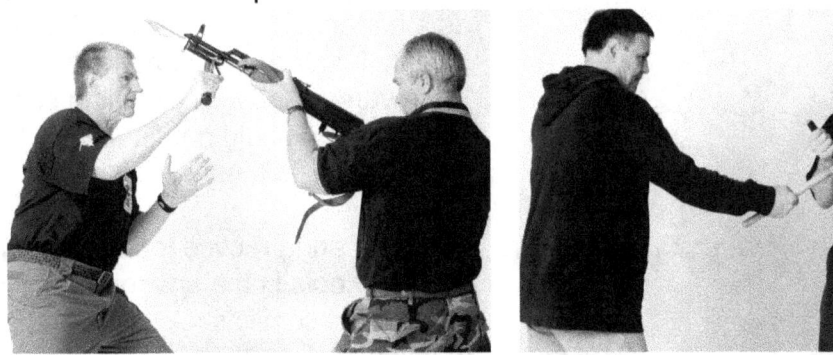

There are saber and reverse grip blocks.

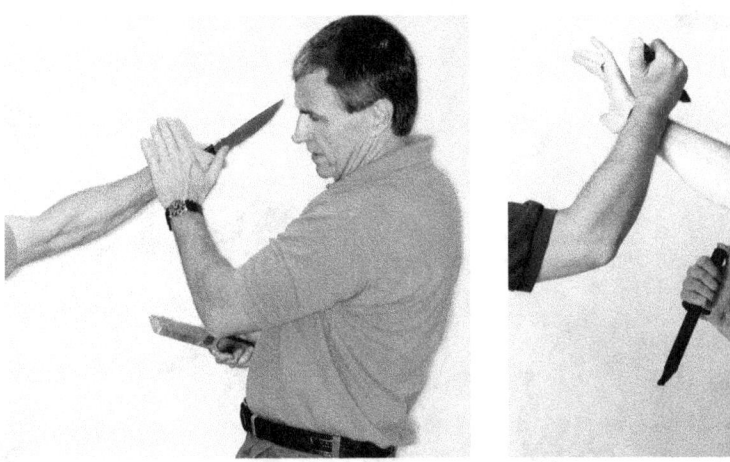

And support limb blocking.

Support Skills – Limb Blocking Basic Training

Here are classic, old school military style support limb blocking. The big blood vessels and finger muscles are turned in.

12 o'clock high.   *3 o'clock right.*   *6 o'clock left.*   *9 o'clock left.*

Never allow yourself to simply knife-to-knife "duel" with the enemy. Whenever possible pick up an object to use as a shield, as a battering ram or as a projectile, or shield. If your empty hand is truly "empty?" Knife versus knife only dueling should be because you could not find something else to grab. This is a universal truth in all combatives.

Learn and practice to identify expedient blocking tools in the environments where you expect conflicts. Use any and all of them in a fight.

Here, I posit the *Minimization of Wounds Theory*. Simply put, rather than have your throat stabbed or slashed, a reflexive forearm block or knife block might save a soldier from a deadly neck wound and leave him with a minor slash on the back of his forearm. It is important to note that this is not a "sacrifice arm" move. We do not wish to sacrifice anything. This word sacrifice is a bad term and reflects the wrong mentality. Instead we want to minimize the wounding, and maximize survival.

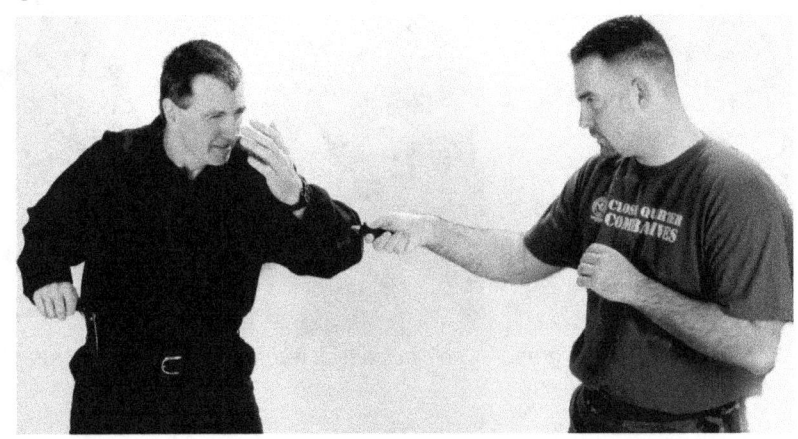

Minimization of Wound Theory. A stab to the forearm is better than a stab to the heart or throat.

Support Skill – Grabbing Skills In Knife Combatives

Your support hand and arm can grab and/or wrap the opponent and help you in your combat situation. Even your weapon bearing limb can capture the enemy's limbs at times. In various military training manuals, the simple act of grabbing the enemy and then stabbing or slashing the enemy is often demonstrated, but usually in sentry killing measures. Rarely if ever is there a competent doctrine that is thorough with a training comprehensive progression.

Remember the prior page on the four arm deliveries. One of the deliveries was the pumping, jabbing attack, whether by thrust or by hook. The pumping attack is quickly pulling away from the target and the grabbing hand, making pumping jabs very hard to grab.

The Knife Combat Grabs and Arm Wrap Grabs Are:
  Grab 1: The enemy's body.
  Grab 2: The enemy's upper arm.
  Grab 3: The enemy's forearm.
  Grab 4: The enemy's wrist.
  Grab 5: The enemy's hand.
  Grab 6: The enemy's impact weapon or firearm from carry to draw.
  Grab 7: The enemy's knife itself in really dire circumstances.
  Grab 8: The knife hand grabs/pinches (usually the enemy wrist).
  Grab 9: The knife-side limb wraps and/or pinches.
  Grab 10: The leg grabs and/or wraps while kneeling or on ground.

Support Skill Basic Grab Solo Practice
Any saber or ice pick knife attack first and then any grab, or...
Any grab attack first and then any saber or ice pick knife attack.

12 o'clock attack with grab.   3 o'clock attack with grab.   6 o'clock attack with grab.   9 o'clock attack with grab.

There are dangers in grabbing gear and clothing. When a soldier seizes the body uniform of the enemy, his clothing, belts, ruck, helmets, web gear, may shift, rip, break or slip. At times the uniform may be a handle and at other times a lost grip and a missed opportunity.

For one example, the defender grabs the attacking arm, and gets a handful of clothes. If the enemy's sleeve is loose, as many are, the knife arm may still slip forward inside the capture and still stab the defender.

Also, the enemy may pull his arm back, as shown to the left. And his arm retracts. This is not a total loss, but may well be an unexpected surprise to the defender, and certainly not the solid grab he for which he'd hoped.

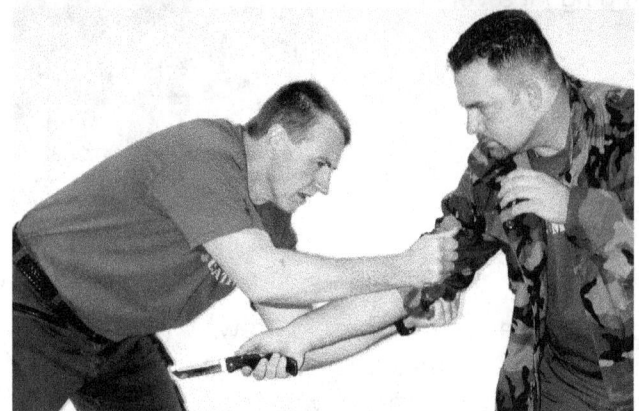

Consider what you are grabbing on the enemy and be prepared for the problems that might occur. When fighting enemies dressed in civilian clothing as well, shirt and jacket buttons may pop. Belts and belt loops may rip. In the case of a bare arm, the enemy's arm may be slippery from blood, or sweat, rain, mud or any number of substances.

The dangers of grabbing the actual knife! In the history of crime and war, people have been forced to grab the blade of the knife itself in close quarter combat. While this is ordinarily ill-advised and a method of very last resort, some people are alive today because they took this action and fought on. And of course, many more are not. Countless military and civilian autopsies note the common "defensive cuts" on the fingers, hands, and forearms of failed grabs.

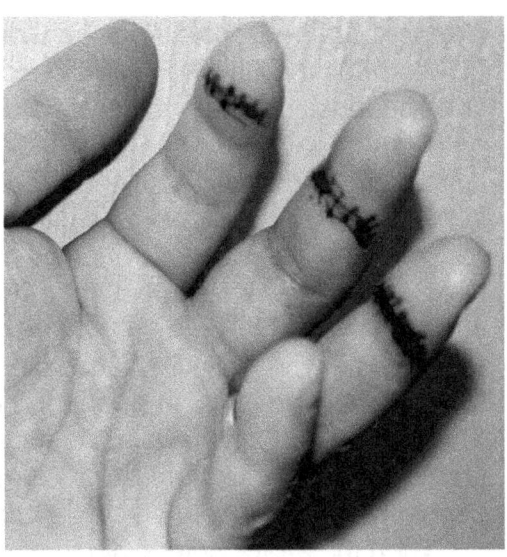

Should an enemy soldier grab your blade, you must pull the knife back and slice your way out of his grip. This is done rather easily. This will often diminish or even end the gripping ability of your opponent by severing important finger muscles.

Once again, I remind practitioners that knife instructors may approach people in practice, when they hold dull training knives, and suddenly snatch the knife to test the intensity of the knife grip. It's a dull knife grip test,

While holding him. Here are samples of a hand grab and an arm wrap.

Common hand grabs.

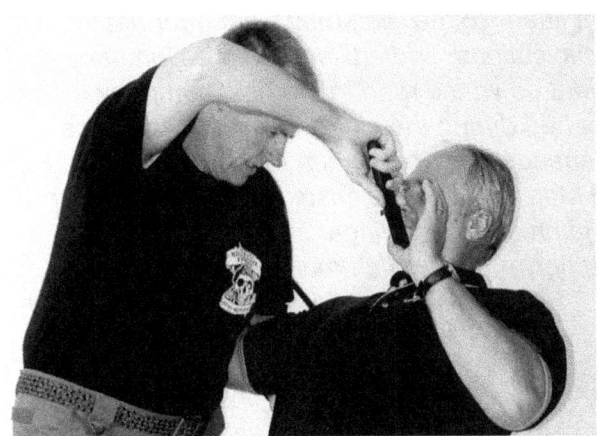
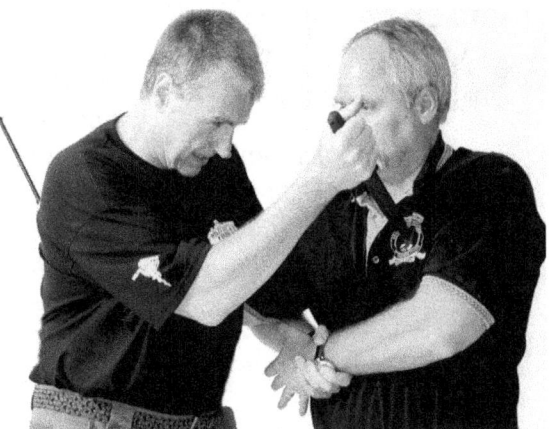

An arm wrap... *...and arm wrap AND a hand grab trap.*

Support Skill – Developing the Grab: Track the Biceps Method

In the 1980s the British Police developed a high-yield, methodology for grabbing the knife-bearing limb of an enemy. The basic premise was not to follow the actual knife hand or the knife wrist with your eyes and your hands. Consider the large circle of travel the knife, hand and wrist has at the end of an arm. The ability to watch and snatch something in that large a circumference is difficult.

Next consider the somewhat smaller circle of travel the weapon-bearing forearm has. Next, the elbow. Smaller, but still a real challenge for the eyes to focus on and the hand to snatch.

The police program instead looked in on the upper body, yet monitored the lower shoulder and biceps of the weapon-bearing limb, using it as a handle to track the arm. They then tried to move with a giant C clamp hand in on the biceps. The biceps and shoulder is easier to monitor, has a much smaller circle of travel and still allows for peripheral vision to see the whole body.

When the attack enters into the C Clamp, seize it like a Venus Fly Trap. Try not to chase and snatch at a moving limb. While grabbing is always hard, the chasing-snatch is the harder. The C clamp grab can then slip down to lower arm grabs.

There will be times and situations when soldiers can track and catch the weapon bearing limb by eyeballing the knife and knife hand. But at times this method may also be used.

Black Box Knife Combat Files

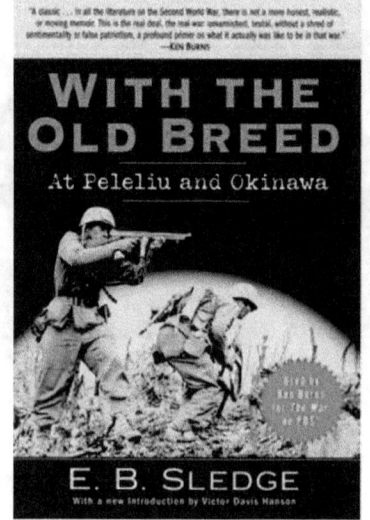

Pelelui, World War II

"The Japanese who had come across the road in front of me (night before) were probably members of what the enemy called a 'close-quarter combat unit.' The enemy soldier shot by Sam was not dressed or equipped like their typical infantryman. Rather he wore only tropical khaki shorts, short-sleeved shirt, and tabi footwear (split-toed, rubber-soled canvas shoes). He carried only his bayonet. Why he entered our line where he did may have been pure accident, or he may have had an eye on our mortar. His comrade angled off toward the right near a machine gun on our flank. Mortars and machine guns were favorite targets for infiltrators on the front lines. To the rear, they went after heavy mortars, communications, and artillery. Before Company K moved out, I went down the road to the next company to see what had happened during the night. I learned that those blood-chilling screams had come from the Japanese I had seen run to the right. He had jumped into a foxhole where he met an alert Marine. In the ensuing struggle each had lost his weapon. The desperate Marine had jammed his forefinger into his enemy's eye socket and killed him. Such was the physical horror and brutish reality of war for us.

– With the Old Breed, E.B. Sledge
Presido Press

Support Skill – Kicking

Kicking can be vital at the right instant. The kicks must be down standing, kneeling (hard for some or most, but try them out!) seated, and certainly on the ground. The *Force Necessary: Hand* course covers these kicks one by one in it's progression. Look there for detailed instruction if you need it. But here, do your best, as none of these kicks are "rocket science."

* Frontal snap kicks.
   Kick the shins. Kick the groin. Use your shoe tips, the tops of your shoes and your shins. He's bent over? Kick any where it hurts. On the ground/floor? Kick anywhere it hurts.

* Stomp kicks.
   Stomp the top of the foot. He's bent over or on the ground? Kick anywhere it hurts.

* Thrust kicks.
   Kick anywhere it hurts.

* Round kicks.
   Kick anywhere it hurts.

* Knees
   Knee anywhere it hurts.

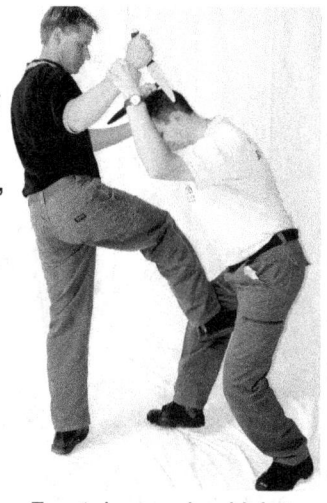

Frontal snapping kick to shin and groin.

Thrust kicks to the leg.

Round kick to legs.

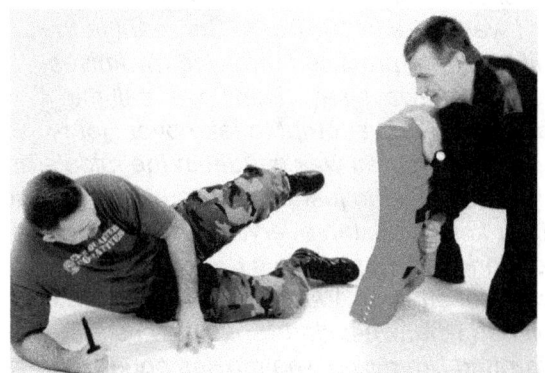

Mandatory! Practice all the kicks on the ground. In the knife course, while holding a knife. Learn to not cut or stab yourself in the process.

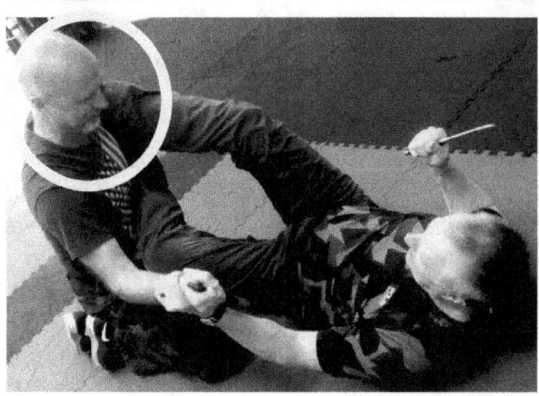

Here's a simulated thrust kick to the head.

Black Box Knife Combat Files

"In the summer of 2007, I was walking around the Animas Park in Farmington, New Mexico. I was wearing my camouflage pants that day, and as I was passing by two older Natives, 6one of the guys called out to me. He asked if I was in the military. I replied, "Yes." He said try throwing these knives. This man had 3 throwing knives next to him on the bench. I said, "What do you want me to throw them at. He said this tree. The tree was old and thick, but I said, "Okay." The first two knives stuck for a second but then fell, and the third one I threw stuck and stayed. With a surprised look on both their faces they smiled back at me.

I said, "Now it's your turn." The one guy took the knives and threw them at the tree. He did as well as I did. We then went back and forth a few more times and most of the throws from both of us were sticking almost every time.

I asked, "What's your name?" He replied, "Justin "Thunder-boy" Morris. I was called Thunder-boy because there was a thunder storm the day I was born. I'm 58 years old now." Both the men were Apacheis de Nabaju (Navajo). I told him I was Lipan Apache and he was happy to hear that. I asked him, "Do you know the foot game." He said, "Yes," and we began to throw the knives around each other's feet without hitting the foot, but getting as close as we could to each other. Again we both did well as no one got a knife thru his foot. I asked him if he knew the Hand-Knife game. He said yes. We both were taught this game at a young age. It's when you stab a knife in between your fingers as fast as you can without cutting yourself.

We talked about knives for a few more minutes and he said, "Snake, in all these years, you are the first person I've ever met that knew how to really throw a knife." As we talked he began to tell me that he was a Vietnam Veteran.

He was at Hamburger Hill, during the war. He said, "Snake, I always carried two knives, two 45 pistols, my M16, and a bright flashlight during the war. I was there in 1967 as a Sergeant in the U.S. Army. I did two tours and spent two years of my life there. I practiced throwing my knives every day. When some of the guys called me "Chief," I would always say, "Don't ever call me "Chief," I'm a Warrior, not a chief!" I was the point man and on one attempt to take over Hamburger Hill, we were losing the battle, and our men began to retreat. I was caught in the middle and so I pulled a dead "Gook" over me and played dead. I did not want to just play dead on the ground, because those "Gooks" would come around and use their knives to stab everyone that was lying down to make sure they were dead. I waited until it got dark and then made it back to my unit. Only seven of us from the front line survived the first wave of attacks. We called that hill names like "Hill 13" and "The Bloodiest Mountain." Hamburger Hill had what we called "Spider Traps." These were holes for point men. The "Gooks" all lived underground like mice. The movies about Hamburger Hill are too much Hollywood. It was far worse in real life. I was with the 101 Airborne and 107 Airborne. The Marines came and helped us out. I was told, "Point" is your life, take care of it. It's "You" or "The Enemy." It's "Yes" or "No." It's "Life" or "Death."

One day, on that bloody hill, my commander told me to take out a machine gunner that kept killing our guys. The machine gunner was well covered with only a small opening for me to make a kill. I snuck up close enough to him without being detected and I got within range to throw my knife. I threw my knife at that "Gook" and it was a perfect shot into his temple. The "Gook" slumped over dead. We call a knife kill a "Solid Kill." Since the kill was also silent, I was able to sneak back to my unit without arousing the enemy. I got a lot more respect after that and nobody messed with me from then on. We eventually overtook Hamburger Hill, but at a huge cost of lives."

– Thunder-boy, Snake Blocker

FORCE NECESSARY: KNIFE!
KNIFE FIGHTING
Knife vs. Hand - Knife vs. Stick - Knife vs. Knife - Knife vs. Gun

Getting Ready, Moving In...
- Quicker Kill, High Yield Targets Study
- The Passing Introduction
- The Ambush Exercise
- The Statue Drill/Exercise
- Counters to Common Blocks Exercise

Skill Developing Drills as the Defender Moves in...

Primer – Beware the Loop!
The army drills, sports team drills, ring fighters drill by isolating steps and working on those sections. The term drill may mean several things in a military framework, from a reservist's weekend, to marching, to a field exercise. To the martial artist, drills come in many different names like sensitivity, synergy, flow, energy and attribute. All are common martial arts monikers for the patterned exercises for two people standing before each other and practicing repetitive moves.

I know many martial artists who are flow-drill and pattern-drill experts. They know and exercise hundreds of choreographed drills, performing them in an artistic, beautiful manner. But many cannot really fight successfully against a chaotic, madman, blitz attacker. This is because they have become drill experts, not fighting experts. They have prioritized the wrong end product of their study.

In your martial close quarter combat drill study you must de-prioritize the drill and re-prioritize the crisis rehearsal of combat scenario practice. Practice drills only support reality. They are not reality. Everyone easily understands this but fails to measure their training clocks when working out, so often dedicating their time to learning forms, katas, and the flow drills and the redundant sets of dead grandmasters.

The Curse of the Loop
The curse of the loop is the real cause. Step A, B, C, D, then A, B, C, D…and so on. These multi-step, practice patterns are meant to be broken with inserted tactics to fire and/or counter. Across the world, martial students are making the grievous mistake of struggling to invent ways to get back into the looping pattern after the inserted tactic. I have seen practitioners toil and sweat to invent ways to get back into the loop of their drill. Many are more proud of the looping drill they invented than the execution of the actual tactic. But the training loop is meant only to be broken. Break the loop with the tactic. Start the loop over.

Real world combat is full of training stories of looping mistakes. One of the worst is of the California police officer that spent many hours disarming a pistol from a training partner. He performed hundreds of repetitions, snatching the gun and quickly handing it back to the partner immediately to keep a training flow going. As you might expect, one day he disarmed a criminal and reflexively handed the handgun right back to the criminal. The criminal shot him dead.

Knife fighting is a lot like football or rugby with a knife. It's sudden and powerful. Filipino weapons expert Remy Presas would often say, "you train your whole life for a 4- second stick fight." As a follow-up he would add, "sometimes you just need one good fake." Like a vicious football tackle or rugby collision, it can all end fast. Keep the *curse of the loop* in mind when you practice. Forget about always rejoining the loop at the end of a string. It is just an unproductive and dangerous mind game. End the loop with your tactic. Stop. Then start again.

Chapter 16: Quicker Kill, High Yield Targets and Attacks

Attacking anywhere on the body with a knife causes damage. It's situational and positional. It has been said by combatives critics that a stab or a slash is not anymore lethal than the other, but rather it matters more *where* you stab or slash. This might seem true in terms of end results but in combat, but we cannot always base our training on the results found on a cold autopsy table. In combat, time matters. It really matters. We do what we do when we have to do it. And, at times in combat we cannot and should not wait for the bleed-out rates of slashes to count an enemy down and out.

Military and criminal forensic science reports that stabbing yields both a higher death rate and perhaps more importantly, a quicker death rate than slashing provides. Stabs are more productive than slashes.

The exercises here cover these high-yield stab and some slashing targets. The exercises in this segment are not meant as a study in interactive, freestyle combat scenarios. This will come later. This is a first-level, lesson in high-yield, high-kill attack targets. The trainer is giving the trainee the most rudimentary attack movements and set-ups so that the soldier can learn, exercise and execute these single, attacks on high yield targets.

These exercises are meant to be simple and quick, two or three set step moves that teach the core, knife combat movements inside more chaotic, multi-move, freestyle scenarios. In this early phase of training the soldier learns the key targeting spots to stab and slash. This will be an progression/evolution into freestyle, combat scenarios.

Quicker Kill Target One: The Brain
  The brain via the face eyes and nose.
  Uppercut under the chin. Need a long thin knife.
  Connections to the brain via the base of the skull.

Quicker Kill Target Two: The Neck
  The windpipe.
  The carotid arteries.
  Connections to the brain via the base of the neck.

Quicker Kill Target Three: The Heart
  Horizontal stab through the rib cage
  Front or back. Back harder.
  From under the rib cage.
  From above the rib cage and down, which is harder.

Quicker Kill Target Four: The Solar Plexus/Diaphragm
Solar plexus up into the diaphragm with a classic hook stab or a thrust that is turned upward.

Quicker Kill Target Five: the Medical Centers
The armpit and pelvis are major muscle, blood and nerve "medical centers" of the body. Stab them.

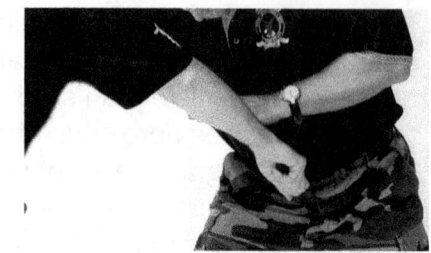

Black Box Knife Combat Files

"I sprang, knife poised ...brought my blade swiftly across his throat, slicing through the flesh and cartilage and grating on deep bone. I thought his head was coming off in my hands. Hot blood gushed over my hands. The stench of it. The stench of the son-of-a-bitch's diet of rotted fish heads with rice, fermented cabbage. I almost retched from the awful stench as I dropped the body and stepped back. It lay pouring blood at my feet, muscles twitching involuntarily. Some of the men dragged the body into the bushes. I didn't look at it again. It was a job. SEALS were paid for doing such jobs. I wiped my blade and my hands on my fatigues."
– Roy Boehm with Charles Sasser,
First Seal,– Pocket Star Books

Black Box Knife Combat Files

"...happened to me once. Two Japs tried with a lariat to hook my foot and pull me from the perimeter. I had the where-with-all to pull out my knife. I cut the rope, stabbed one Jap right in the eye, and slashed the other. ..got free, then got back in the foxhole. Then, I rolled a hand grenade over the edge and yelled, "Crooks! Grenade! To get me buddies to duck."

– Bill Crooks, Australian Army,
Oral History of South
Pacific Theater

Black Box Knife Combat Files

"Suddenly, a pistol cracks in the room. A puff of gunsmoke rolls over us. The bullet hits the wall in front of me. Where did that come from? Does he have a sidearm? I cuff him across the face with my torn left hand. He rides the blow and somehow breaks my choke hold on him. I bludgeon his face. He tears at mine. I gouge his left eye with my right index finger. I am astonished to discover that the human eye is not so much a firm ball as a soft, pliable sack. He wails like a child. It unnerves me, and I lose the stomach for this dirty trick. I withdraw my finger. Something metallic hits the cold concrete flooring. As he reaches for his [dropped] pistol, I slam my left fist as hard as I can down onto his collarbone. He swings wildly at me again. My helmet's gone now. I have no idea where my M16 is. I feel my strength ebbing. I don't have much left. He kicks at me, throwing his whole body into it. I've got to end this. But I don't know how. "Surrender!" I'm gored. He fights on, and I can sense he's encouraged and he's close to getting free of me. I swallow hard and gag.

My mouth is full of blood, and I don't know whose. Somebody shouts something. I listen for Arabic. I think I hear, "Are you okay?" and "God!" The man beneath me tries to answer but I cork him with another fist to his face. He takes it and jabs weakly back at me. Blood sprays from his face and speckles onto mine. My grip on him loosens. One more push, and he'll be free.

I have a knife on my belt. I sit up, putting my weight onto his chest. Slowly I get to my feet. My legs are spread, my center of gravity low. I reach for my belt just as he comes up after me. His face rams my crotch. I feel his teeth clamp onto me. Oh Fuck. I pummel down on his head, but he grinds his teeth harder. Searing agony, pain...It takes a monumental effort to unhitch the Gerber from my belt. I use it as a bludgeon. At first, my blows are pathetic. They land on his head and do nothing to dissuade him. He growls and screams and holds down his bite. I'm almost paralyzed with the pain...Finally, suddenly, I become a madman.

My arm comes up over my head, then chops down with every bit of power I have sends the Gerber's handle thundering down on the enemy's head. He sags back onto the floor. I can feel warm liquid trickling from my crotch down my legs...I flick the Gerber open. The blade locks in place.

I pounce on him. My body splays over his and I drive the knife right under his collarbone. My first thrust hits solid meat. The blade stops, and my hand slips off the handle and slides down the blade, slicing my pinkie finger. I grab the handle again and squeeze it hard. The blade sinks into him, and he wails with terror and pain. The blade finally sinks all the way to the handle. I push and thrust it, hoping to get it under the collarbone and sever an artery in his neck. He fights, but I can feel he's weakening by the second. I lunge at him, putting all my weight behind the blade...His mouth is curled into a grimace. His teeth are bared. It reminds me of the dogs I'd seen the day before. The knife finally nicks an artery. We both hear a soft liquidy spurting sound. He tries to look down, but I've pinned him with the weight of my own body. My torn left hand has a killer's grip on his forehead. He can't move. I'm bathed in warmth from neck to chest. I know it is his blood. His eyes lose their luster...he takes a last breath and his eyes go dim, still staring into mine."

– SSG David Bellavia, *House to House*, Free Press

Chapter 17: The Knife Passing Introduction

You cannot pass a "hit and retract attack. There is nothing to pass!"

Review the definitions of the definitions of the words ""deflect," and "pass." Passing a knife attack means getting it from one side to the other. High to low. Low to high. Right to left. Left to right. Review the four arm deliveries – the lunge, pumping hit and retract, the thrust and the hook. You'll see you can only pass committed, hooking attacks.

Technically, you block/deflect thrusts. You cannot pass any pump/jabs attacks because the incoming, continuing, passable energy has stopped and is retracted back away. There is simply nothing left available to pass across.

Both the empty hand limb and the knife-bearing limb can connect with the attack, pass and hooking style, committed energy, attack, be it knife or empty-hand.

The following training segment list will cover knife versus a hooking knife that is committed to lunge. These drills are for passing only. Pass and counter-attack combinations training will appear later in this manual in the combat scenarios chapters.

The following checklist is the most comprehensive collection of passing exercises.

Basic Passing Set 1: Empty Hand - Pass the Slash Attack
Empty hand pass of a 12 o'clock saber slash hook attack.
Empty hand pass of a 3 o'clock saber slash hook attack.
Empty hand pass of a 6 o'clock saber slash hook attack.
Empty hand pass of a 9 o'clock saber slash hook attack.
Empty hand pass of a 12 o'clock reverse grip hook attack.
Empty hand pass of a 3 o'clock reverse grip hook attack.
Empty hand pass of a 6 o'clock reverse grip hook attack.
Empty hand pass of a 9 o'clock reverse grip hook attack.

Basic Passing Set 2: Saber Grip - Pass the Slash Attack
Saber grip pass of a 12 o'clock saber slash hook attack.
Saber grip pass of a 3 o'clock saber slash hook attack.
Saber grip pass of a 6 o'clock saber slash hook attack.
Saber grip pass of a 9 o'clock saber slash hook attack.
Saber grip pass of a 12 o'clock reverse grip hook attack.
Saber grip pass of a 3 o'clock reverse grip hook attack.
Saber grip pass of a 6 o'clock reverse grip hook attack.
Saber grip pass of a 9 o'clock reverse grip hook attack.
Saber grip switch to other knife hand and cycle through the set.
Saber grip pass with both the empty hand and saber grip knife together.

Basic Passing Set 3: Reverse Grip - Pass the Slash Attack

Reverse grip pass of a 12 o'clock saber slash hook attack.
Reverse grip pass of a 3 o'clock saber slash hook attack.
Reverse grip pass of a 6 o'clock saber slash hook attack.
Reverse grip pass of a 9 o'clock saber slash hook attack.
Reverse grip pass of a 12 o'clock reverse grip hook attack.
Reverse grip pass of a 3 o'clock reverse grip hook attack.
Reverse grip pass of a 6 o'clock reverse grip hook attack.
Reverse grip pass of a 9 o'clock reverse grip hook attack.
Switch grip to other knife hand and cycle through the above.
Pass with both the empty hand and reverse grip knife together.

Sample: Pass Set 3: You pass-across the committed, hooking power attacks with your reverse grip knife.

REMEMBER! *You cannot pass a "hit and retract!"*

CHAPTER 18: THE KNIFE AMBUSH, DODGE, EVASION EXERCISE

This is ambush and respond exercise. At times, a criminal or a soldier approaches you with a grin and a story. He gets close. We can never replicate in a training setting, a true ambush, but we try with these 10 situations. A trainer stands before you. Close, but not too close. Your hands are down because this is an ambush. This is a step-by-step isolation skill developer. He suddenly attacks!

1: A right hand, high, hooking strike from his hight right.
2: A right hand, high, back-handed, hooking strike from his right hand.
3: A right hand, belly high, hooking strike from his hight right.
4: A right hand, belly high, back-handed, hooking strike from his right hand.
5: A right hand, thigh high, hooking strike from his hight right.
6: A right hand, thigh high, back-handed, hooking strike from his right hand.
7: A right hand, low to high hook.
8: A right hand, hight to low hook.
9: A right hand thrust to the stomach.
10: A right hand thrust to the face.
* Reset, and attack with the left hand.
* Attack with hand, stick and knife.

Note and remember - each weapon stab and slash attack is an individual ambush attack. This represents ten ambush attacks. The attack can come from a stick, bat or edged weapon, something to justify a deadly force knife response.

The trainee will be asked to respond in a myriad of ways, to dodge without footwork, Then add footwork. Then block. Then grab. Then pass and draw. Then grab and draw. Then add footwork and do all this. Then do this on the ground versus a standing attacker. Then while being held in a bear hug from behind. Its original main purpose is to create "dodging flexibility."

Ambush attack 1: Head shot from trainer's right. An inward strike.

Ambush attack 2: Head shot from trainer's left. A backhand strike.

Ambush attack 3: Belly shot from trainer's right. An inward strike.

Ambush attack 4: Belly shot from trainer's left. A backhand strike.

Ambush attack 5: Knee shot from trainer's right. An inward strike.

Ambush attack 6: Knee shot from trainer's left. A backhand strike.

Ambush attack 7: Shot up from trainer's low. Groin target? An upward strike.

Ambush attack 8: Head or clavicle shot from trainer's high. A downward strike.

Ambush attack 9: Belly stab. *Ambush attack 10: Face stab.*

Isolation Set 1: The First Response - Body Dodging.

This first set, works on an officer's flexibility and elasticity. In an ambush, it is likely the first response will be a body dodge. No footwork yet! Except a little in-and-out footwork, versus the lower leg attacks. In the exercise, the trainer attacks with a stick for a series, with a knife, and then empty handed.

Dodge versus 1 and 2: Save the head/neck.
  * Chin pull, or –
  * Bob and weave, or –
  * Duck.

1: Inward head slash. *2: Backhand head slash.*

Dodge versus 3 and 4: Save the stomach.
  * Hollow out torso (hula-hoop).

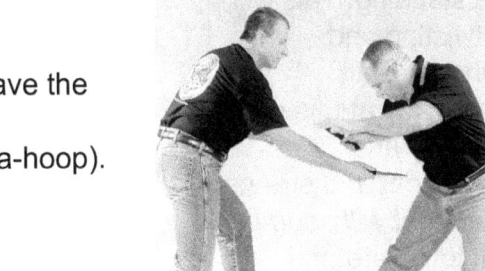 

3: Inward belly slash. *4: Backhand belly slash.*

Dodge versus 5 and 6: Save the knee.
  * Step away or *really* twist the knee. Stepping away and back is smarter, but some do teach the knee twist.

5: Inward knee slash. *6: Backhand knee slash.*

Dodge 7 and 8: Save the torso
   * Twist torso

Dodge 9 and 10: Save the torso and the head.
   * Twist torso, or
   * Hollow out torso
   * Bob, weave, or,
   * Slip.

7: Upward slash. 8: Downward slash.

9: Stomach/guts stab. 10: Face stab.

The Rattlesnake Dodge
The trainer attacks with a knife using these 10 angles. The trainee does his best to dodge and evade the attacks. The trainer will not be able to use the same skills as when standing. This will involve rolling, "shrimping," twisting and whatever it takes to dodge the assault.

This is just an elasticity exercise at this early stage, to create evasion elasticity and maneuvers on the floors, pavement and streets. Blocking with arms and shoed feet *will soon be added later* as it is important to isolate and develop the evasion skills alone.

Remember each attack is supposed to represent one ambush attack. The officer is knocked down and is attacked by a hand, a stick, or a knife.

(To end each set, the trainee may kick at the trainer's legs to back him off and away, and then practice getting up to his or her feet as fast as possible.)

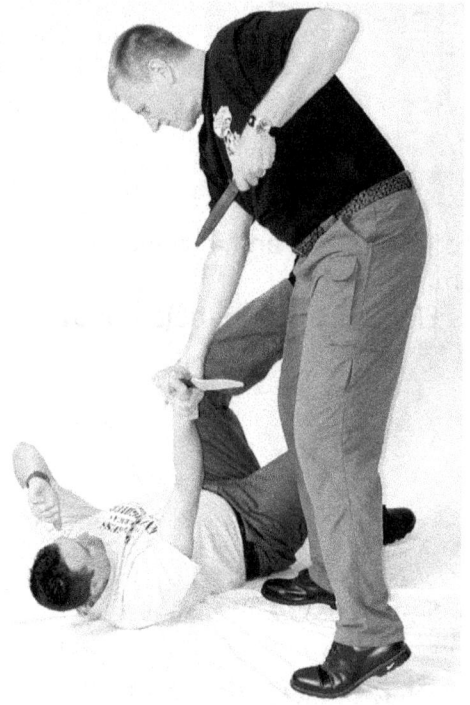

You will experience the fact that Ground/dodging is a new skill.

Two Ambushers
In order to further develop the spontaneity of a sudden attack, ambush skills, two trainers stand near the trainee. In an unplanned, surprise manner, both trainers suddenly, and separately attack the trainee. They do not take turns, but rather break patterns. This heightens the surprise attack detection and develops various awareness attributes of the trainee.

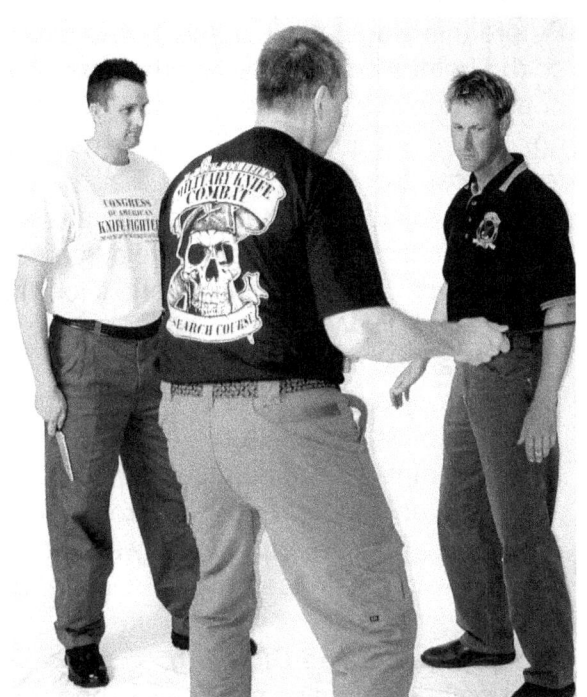

Dodging and Evading While Being Held
One trainer holds the trainee from behind in a rear bear hug while the other trainer strikes at the trainer. Under this pressure, the trainee tries to dodge and evade the strikes. This requires extra effort, and the trainee usually can move the trainer holding him, providing the trainee is not unusually strong or large in comparison to the trainee. Experiencing this predicament will help define this analysis. By moving this extra weight, the trainee also develops better evasion motions when he or she is unheld.

    Set 1: The trainee is attacked by empty hand strikes and kicks
    Set 2: The trainee is attacked by a knife

If the trainees are well training and experienced in combatives, at the end of each set, trainees might finish the set by escaping from this rear bear hug. The trainee can simulate foot stomps, attacks to the groin, torso pivots, or any number of common self defense escapes from this bear hug hold.

Isolated Set 2: Add Footwork

Before this stage, this set, the trainer did not move in, so the trainee could develop body dodging skills. The body dodge in an ambush may be all the defender can do. Now, the trainer can move in and around, and the trainee can too. Think about football, basketball and soccer. there is an endless amount of footwork. Throw your "athletic" switch. Begin a lifelong study of agility and footwork. Obviously, footwork is for standing only.

> Sample: Walking.
> Sample: Running.
> Sample: In and out.
> Sample: Shuffle foot.
> Sample: Grounded. Get away and get up.
> Note! Once footwork is introduced, we can add the X slashes. Body elasticity alone will not dodge these X angle attacks.

Isolated Set 3: Add Blocking

In an ambush, the first blocks are done unarmed. The trainee gets to exercise blocking skills versus the attacks. In doing so, the trainee will still use body evasion skills and footwork, but now will also block. It is also reflexive to block. The trainee should try to block on the weapon-bearing limb forearm of the trainer, not on the weapon itself, for obvious reasons. The trainee should avoid bending over and blocking any low-line kicks to their legs, using lead leg avoidance instead. Essentially there are two kinds of blocking, methods versus unarmed attacks which can resemble MMA kick boxing (closer to the body) and methods versus weapons (perhaps further from the body). Begin a lifelong study of unarmed and mixed weapon blocking.

When blocking versus a knife, experts suggest turning your forearm inward to protect the soft tissue, muscles and blood veins.
Note: *The support body twist developed by working these body evasion exercises.*

> Review blocking on the Combat Clock
> – single arm series.
> – double arm series.
>   * parallel forearms.
>   * doomsday and vision concerns.
>   * high-low, detached X.
>
> – versus unarmed ambush attacks.
> – versus stick ambush attacks.
> – versus knife ambush attacks.
> – standing.
> – grounded. Pivoting on the small or your back, or hip and blocking with your shoes, versus unarmed, stick and knife attacks.

Once again, use your shoes or boots and hip pivots to help block.

Isolated Set 4: Add Grabbing with Hands or Arms

You might have a good grab but a bad grip? You might have a bad grab and a good grip. Obviously it would be wise to grab the limb and not the stick or knife. It can be grabbed in three phases.

- before the strikes lands.
- maybe somewhere it seems to be landing.
- after the dodge and it passes.

Begin a lifelong study of unarmed and mixed weapon grabbing and gripping. Exercise and isolate grabbing versus the Big Ten attacks.

- versus unarmed ambush attacks.
- versus stick ambush attacks.
- versus knife ambush attacks.
- standing - **Note:** *no grabbing of the leg kicks.*

– grounded, versus unarmed, stick and knife attacks.

Isolated Set 5: Add Knife Quick Draws

Do what you must to dodge and evade the attacks. Grab or don't grab, but practice your stress knife quick draw under this weapon attack.

You can add counter attacks now, official as our progression continues. You are now closer in and probably dealing with forearm contact, leading us to the next series of engagements.

Chapter 19: The Statue Drill Exercises

Body contact may be a completely new experience for most defenders entering martial studies. Many martial arts use this method of training to introduce the newest practitioners to the initial stages of body contact. We call it the statue drill and it resembles an affordable Wing Chung dummy, as the wooden dummies may cost hundreds if not thousands! Your training partner is free.

Plus, this drill develops a full spectrum of experimentation, covering attack options often missed without this approach. The trainer stands in a statue for the trainee to exercise hand, stick and knife strikes. The trainer may stand in 7 different spots to best develop the trainee.

A reference arm contact point is where the trainer and trainee meet arms in very common close quarters combat contact. In stand-up combat and at times in even knee-high, seated and ground combat. The complete formula for arm contact reference points work straight across the statue from the trainer's right to left.

It is worth mentioning from the start, that the list has two very, what some martialists call, high-level training contacts, the so-called "splits" or the two split-arm situations, where the trainee ends up with both his arms on both sides of the trainer's right or left arm. Including these split arms might be considered quite advanced training, even unnecessary by some standards, while the basic training list, minus the splits is certainly quite a competent and adequate exercise.

These 7 positions/spots/situations/encounters are:

Spot 1: Outside his right arm.
Spot 2: Your "split arms" as in one of your arms outside his right arm, one inside his arm.
Spot 3: Inside his right arm.
Spot 4: Inside both his arms, sort of chest facing chest.
Spot 5: Inside his left arm.
Spot 6: Your "split arms" as in one of your arms inside his left arm, one outside his arm.
Spot 7: Outside his left arm.

Note: In physical training I often do not do the split arms as a high priority. I show them so that practitioners know they exist, but do not work them over the simple...

– Outside left.
– Inside left.
– Inside left and right.
– Inside left.
– Outside left.

Here is the common statue drill position, that allows a trainee to circumnavigate a body and experiment with invasion.

It should be noted that the "statue' can be positioned in any needed structure for the training mission. The trainer stands with:
- a) both arms up.
- b) one up, one down.
- c) both down.
- d) a single step attack pose.
- e) The statue can pump arms.

The Saber Grip Slash Statue Series
1: Knife limb makes contact outside, inside, split, inside, outside the arms.
- a) slash the trainer's neck.
- b) slash the trainer's limb.

2: Support limb makes contact outside, inside, split, inside, outside the arms.
- a) slash the trainer's neck.
- b) slash the trainer's limb.

3: Knife limb makes contact, outside, inside, split, inside, outside the arms.
- a) support hand strikes face.

The Reverse Grip Slash Statue Series
1: Knife limb makes contact outside, inside, split, inside, outside the arms.
- a) slash the trainer's neck.
- b) slash the trainer's limb.

2: Support limb makes contact outside, inside, split, inside, outside the arms.
- a) slash the trainer's neck.
- b) slash the trainer's limb.

3: Knife limb makes contact, outside, inside, split, inside, outside the arms.
- a) support hand strikes face.

The Saber Grip Stab Statue Series
1: Knife limb makes contact outside, inside, split, inside, outside the arms.
- a) stabs the trainer's neck.
- b) stabs the trainer's limb.

2: Support limb makes contact outside, inside, split, inside, outside the arms.
- a) stabs the trainer's neck.
- b) stabs the trainer's limb.

3: Knife limb makes contact, outside, inside, split, inside, outside the arms.
- a) support hand strikes face.

The Reverse Grip Stab Statue Series
1: Knife limb makes contact outside, inside, split, inside, outside the arms.
- a) stabs the trainer's neck.
- b) stabs the trainer's limb.

2: Support limb makes contact outside, inside, split, inside, outside the arms.
- a) stabs the trainer's neck.
- b) stabs the trainer's limb.

3: Knife limb makes contact, outside, inside, split, inside, outside the arms.
- a) support hand strikes face.

Statue Drill Sample Series: The Reverse Grip Stab - knife arm contact and knife slash or stab

Step 1: Trainee knife contact outside the right arm.

Step 2: Trainee slips-cuts knife up and stabs. Cover hand at the ready.

Step 3: Trainee knife contact inside the right arm.

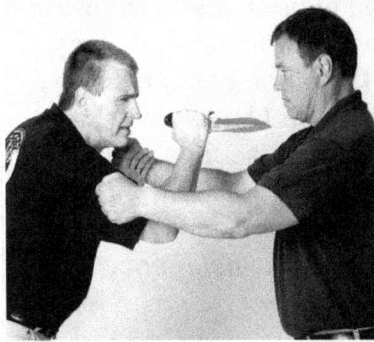

Step 4: Trainee slips knife up and stabs. Cover hand at the ready.

Step 5: Trainee splits contact inside the two statue arms.

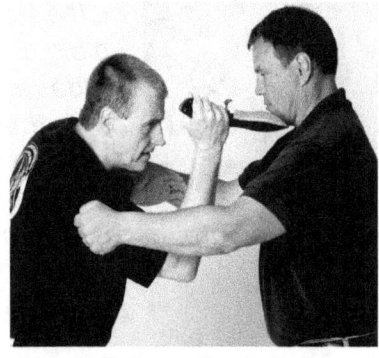

Step 6: Trainee slips knife up and stabs. Cover hand at the ready.

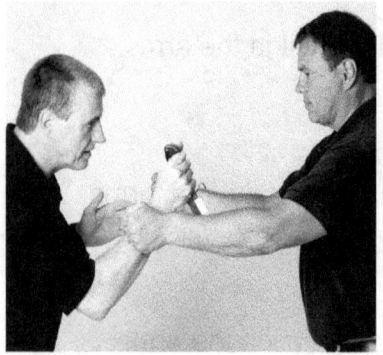
Step 7: Trainee knife contact inside the left arm.

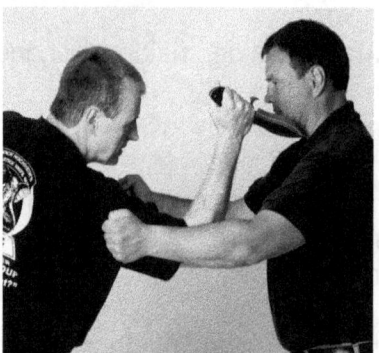
Step 8: Trainee slips knife up and stabs. Cover hand at the ready.

Step 9: Trainee knife contact outside the left arm.

Step 10: Trainee slips knife up and stabs. Cover hand at the ready.

The Statue Training Progression Summary
Contact 1: outside the right arm, then knife attack.
Contact 2: inside the right arm. then knife attack.
Contact 3: split arms, then knife attack.
Contact 4: inside the left arm, then knife attack.
Contact 5: outside the left arm, then knife attack.

Note: remain free and athletic with footwork.
Note: keep cover hand up in the ready position.
Note: knife limb first contact and attack.
Note: support limb first contact and attack.

Black Box Knife Combat Files

The first thing they did was to throw a brace into the wheels of the tank. They stopped it and then swarmed over it like monkeys. I could see one on the tank with a pitchfork. He was jamming it down the turret, trying to get at the fellows inside. There were others brandishing long knives. Then they got some gasoline and threw it on the tank and set it afire. The tank started to burn. It was throwing up big clouds of smoke, and I began to wonder whether the guys inside really were dead or just supermen. I saw one of them come out ... then another came out. Gosh! In all my life I have never seen one man take such a beating. They kicked him in the face and stomach. One pulled off his ear, and another smashed him with fists. One grabbed him with the pitchfork, and another man stuck a knife in him ... "
– " Alex Alan Reynolds, *Insurgency!*
Life Magazine

CHAPTER 20: COUNTERS TO COMMON BLOCKS EXERCISE

This drill recognizes and develops fighting skills at the forearm-to-forearm range. It contains four Combat Clock corner events as a basis and develops counters to common blocks. The trainer acts like a moving training dummy. The trainer may or may not have a weapon in his hands, depending upon the purpose of the drill. When first attempting these drills, it is advisable that the trainer is unarmed as the presence of the knife in the trainer's hands is confusing and initial skill development may be confusing. Later, weapons can be introduced. This builds coordination, speed, target acquisition. These drills are best learned under the watchful eye of a veteran instructor.

Counters to Common Block Event 1: Trainer attacks high. Trainee blocks.
Counters to Common Block Event 2: Trainer attacks from the right side. Trainee blocks.
Counters to Common Block Event 3: Trainer attacks from below. Trainee blocks.
Counters to Common Block Event 4: Trainer attacks from the left side. Trainee blocks.

12 o'clock high strike and block.

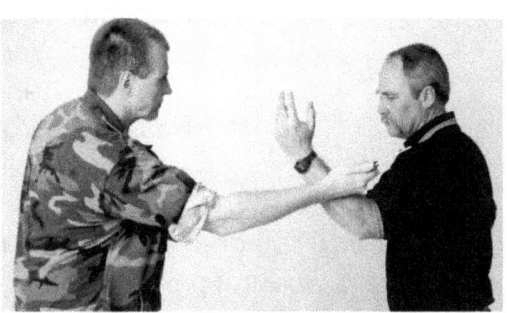
3 o'clock block and strike.

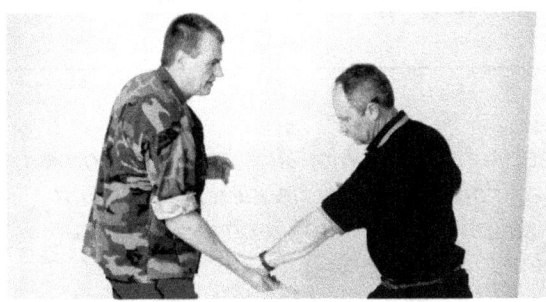
6 o'clock high strike and block.

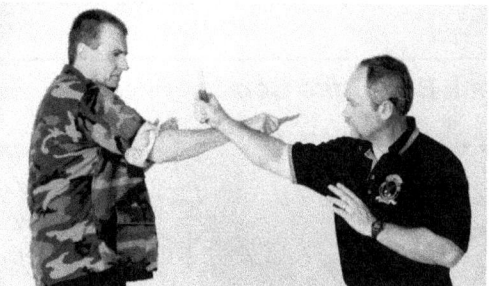
9 o'clock block and strike.

It should be noted that the contact is a reference point and theoretically, it does not matter who the attacker and who the defender is. On contact, it becomes a reference point, high or low, right or left. The contact reference points are:
  Option 1: Knife (or any weapon) block to knife attack contact.
  Option 2: Forearm block to knife attack contact.
  Option 3: Forearm to forearm contact.

Counter 1: Cut the block
The block and strike contact is made and the trainee quickly maneuvers into cutting or stabbing the trainer's blocking limb.

Counter 2: The trainee redirects the attack on another line
Since we operate under the principle of the "uncommitted stash or stab, we can quickly redirect our attack to a wiser and more open target.

Counter 3: Knife invading hands
Or some call, "knife trapping hands." All trapping hands fall under the categories the four "P's", the pinning, passing pulling or pushing of the opponent's limbs to clear a path to a better or best target. The Four Ps are:

P1: Pinning. You pin the opponent's limb by pushing it against himself or a wall or the ground/floor.

P2: Passing. You pass the attack if you feel the consistant energy of the attack, NOT a hit an retract. You pass the attack by you.
This is done two ways.
Passing A: Force to Force-Surrender the Force.
You block, yet feel the incoming pressure, and you manipulate it into a pass.
Passing B: You see the force is incoming and you blend/mix with the energy and pass it by.

P3: Pulling. You grab and pull the opponent's limb out of the way, so as to clear a path to a better or best target.

P4: Pushing. You push the opponent's limb away so as to clear a path to a better target. The limb is not pinned as in P1. Just pushed.

Black Box Knife Combat Files

"We needed some of the ammo, and there was some Jerry explosives right there. We needed it badly, you know. A Jerry sat on a chair in the archway, and another one was puffing away on a fag about 10 meters away, in the dark, his back to me, too. Walking around. I thought if I could stab the Kraut in the chair, simple, maybe I could run up and stab the other too. I got up to the bloke in the chair. It was about 4 a.m. you know, the best time to work such things. He may have been asleep. I grabbed his mouth and stuck me' dagger into his throat. I tipped the chair back, and he died. Just died right there. But I had to twist the knife like screwing it with a screwdriver, ya' know? Just like that. I laid the chair back on the ground. He sat there. Sideways. The other Jerry with the fag just wandered off! Never even heard us. I cut some of the bands on the boxes that was holding the gear, with that very same bloody dagger, and off I went with some."

– Sidney Lawrence, *Commando Combat, Lithglow*

CHAPTER 21: THE KNIFE OUTSIDE INVASION SERIES

Practicing this carefully constructed set will develop skill sets needed to clear the enemy's limbs and invade in for further action. This format is based on developing speed, skill and flow in an outside arm-to-outside arm encounter, right or left sides. Once this reference point is established, the basic training drill steps are:

Counter 1: The knife hits target unblocked and hits the weapon-bearing limb.
Counter 2: The enemy blocks half way and limb is cleared, target hit & weapon limb hit.
Counter 3: Enemy grabs. Releasing technique. Target hit. Weapon limb hit.
Counter 4: Enemy over-blocks. Block passed. Target hit. Weapon limb hit

The High Outside Reference Points

These three, high, outside reference points are obtained by combat contact. It does not matter who is attacking or who is blocking, just that contact has been made. The counters begin from these points.

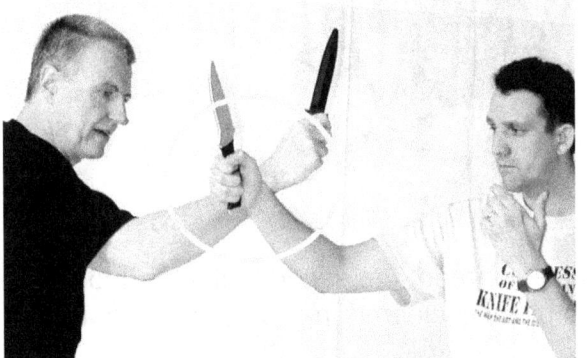
High Reference Point 1: Forearm to forearm.

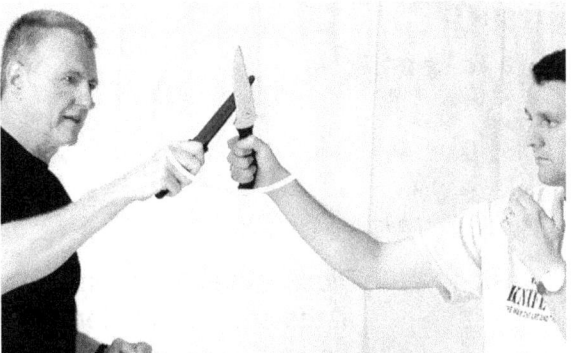
High Reference Point 2: Knife to knife.

High Reference Point 3: The painful knife to forearm, either person.

The following photo series is a sample of these Big 4 situations. They should be done right-handed versus right-handed. Left-handed versus left-handed. Saber grip and reverse gip.

Invasion Set 1: No second block: First obstruction cleared with a push, knife hits a target, then knife hits weapon-bearing limb to hope for an impact disarm.

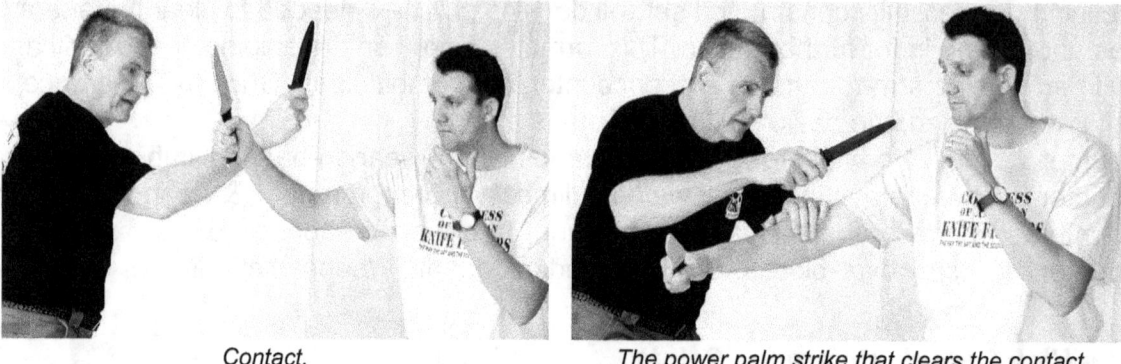

Contact.   The power palm strike that clears the contact.

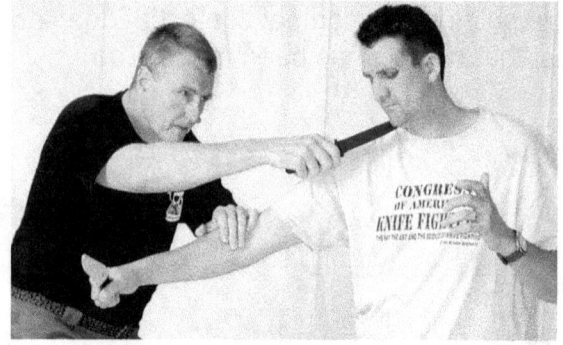 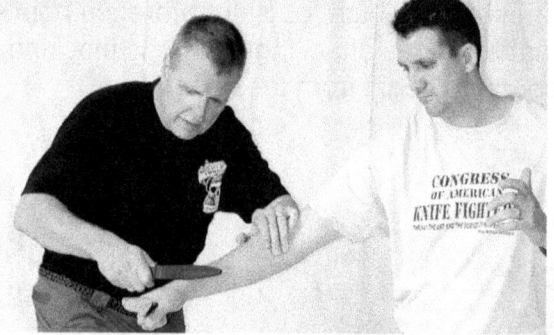

The knife hits a major target.   The weapon-bearing limb is struck.

Invasion Set 2: Blocked half-way: First obstruction cleared, the second attack is blocked at the enemy's center-line. This block is cleared, hits a target, knife hits weapon-bearing limb.

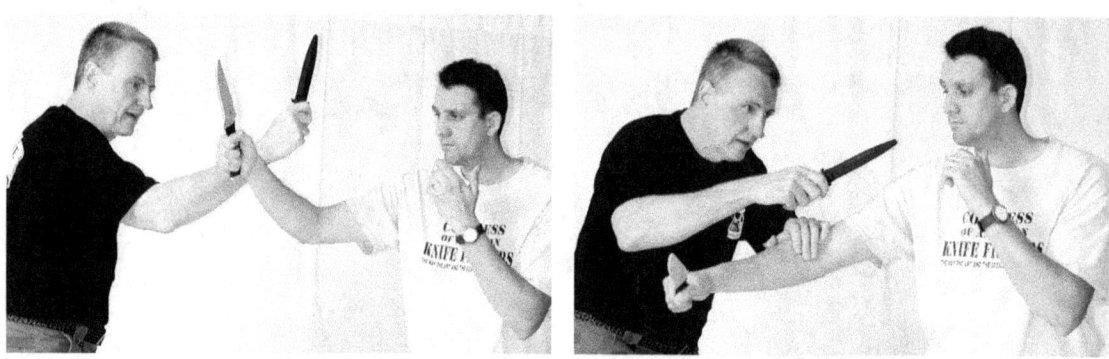

Contact.   The power palm strike that clears the contact.

The knife attempts a strike, but is blocked half-way by the enemy. The soldier's elbow strikes the enemy's right arm with a power torso shift/twist, and the hand shoves the block aside.

The knife strikes a main target.

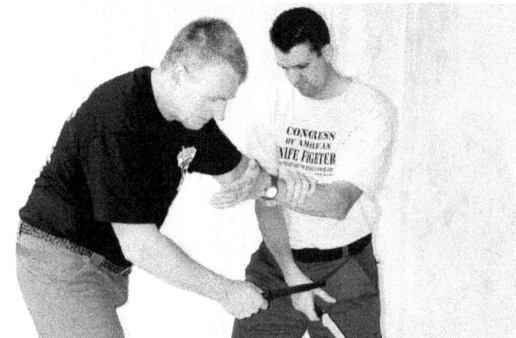
The weapon-bearing is struck.

Invasion Set 3: Attack grabbed: **First obstruction cleared, knife hits a target, knife hits weapon-bearing limb.**

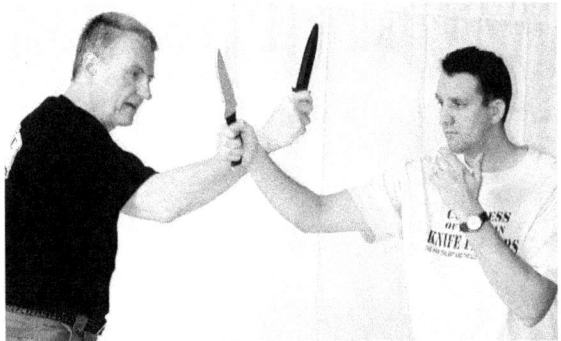
Contact.

The power palm strike that clears the contact.

The attack is caught! Trainee raises his caught arm elbow and shifts it up and over the trainer's catching arm. Trainee drives his elbow downward and frees himself from the grip.

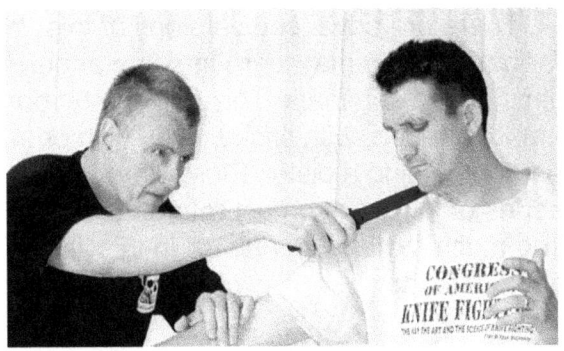
Trainee attacks main target.

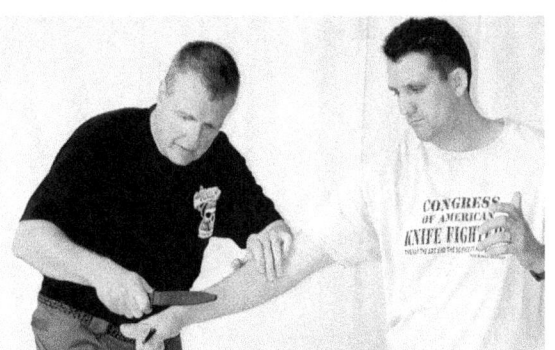
Trainee finishes with a strike to the trainer's weapon bearing limb.

Invasion Set 4: The attack is over blocked (passed his center-line.)

Contact.

The power palm strike that clears the contact.

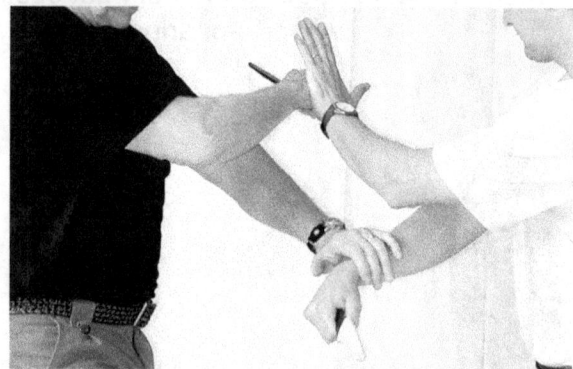
The trainer, in anticipation of serious incoming danger, over-blocks, passing his body's center-line.

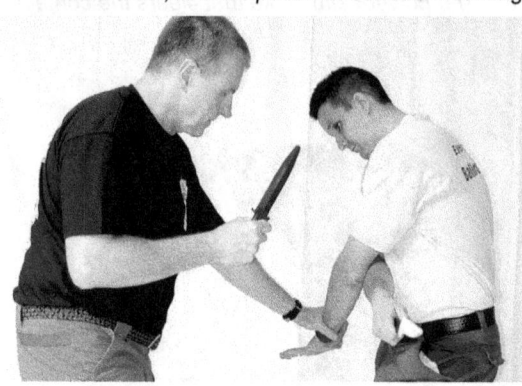
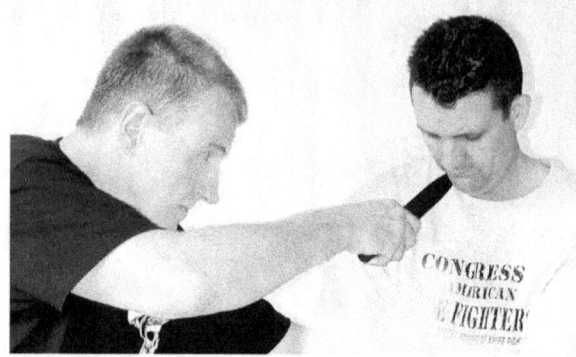
The trainee takes his left hand off the trainer's weapon-limb and quickly grabs the over-extended block, pushing the arm atop the weapon-bearing limb.

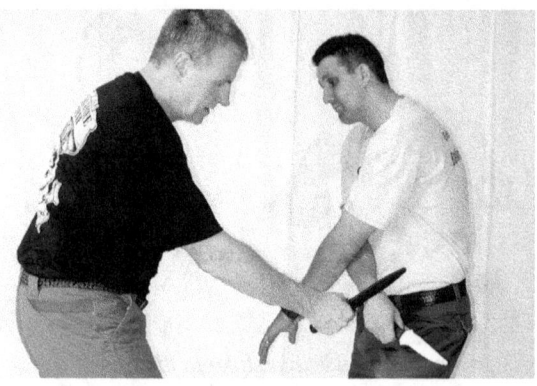
Then the trainee attacks the face or throat of the trainer, then hits the weapon-bearing.

"There are advanced versions of this drill format that you may learn in some of our films and in seminars. This drill can be done empty-handed, with sticks, even with one person holding a pistol. Plus, there is a series of low arm contact reference points. Too much material to publish in this limited subject matter."

– Hock

Chapter 22: The Block, Pass and Pin Exercise

This drill recognizes and develops fighting skills at the forearm-to-forearm range. It contains six events as a basis and develops a multitude of tactical interruptions. The trainer acts like a moving training dummy. The trainer may or may not have a weapon in his hands, depending upon the purpose of the drill. This builds coordination, speed, target acquisition.

Event 1: The trainer strikes downward at an angle. Trainee blocks.
Event 2: The trainee's other arm passes under the arms and pushes off the trainer's arm.
Event 3: The trainee pushes the trainer's arm down in a pinning fashion.
Event 4: The trainee attacks the trainer with a downward strike at an angle. Trainer blocks.
Event 5: The trainer's other arm passes under the arms and pushes off the trainee's arm.
Event 6: The trainer pushes the trainee's arm down in a pinning fashion.
Event 7, 8, 9:...Repeat...

1: Trainer strikes. Trainee blocks.

2: Trainee passes.

3: Trainee pins.

4: The trainee attacks high.

5: The trainer passes.

6: The trainer pins.

1/2 Beat Inserts and Interruptions
At 1 1/2, 2 1/2, 3 1/2, 4 1/2, 5 1/2 and 6 1/2 exercise these attacks.

* half-beat saber slashs to forearm, upper arm, neck.
* half-beat reverse grip slashes to forearm. upper arm, neck.
* half-beat saber grip stabs.
* half-beat reverse grip stabs.
* half-beat pommel strikes.
* half-beat shin kicks.
* half-beat support hand grabs.
* half-beat support hand eye attacks.
* half-beat shoves and knife quick draws.
* experiment with arm wraps, bars and takedowns.
* continue to invent insert 1/2 beat options.

Grounded Block, Pass, Pin Examples

Event 1: The trainer kicks downward at an angle. Trainee blocks.
Event 2: The trainee's other arm passes under the legs and pushes off the trainer's leg.
Event 3: The trainee pushes the trainer's leg down in a pinning fashion.
Event 4: The trainee attacks the trainer with a downward strike at an angle. Trainer blocks.
Event 5: The trainer's other leg passes under the leg and pushes off the trainee's arm.
Event 6: The trainer pushes the trainee's arm down in a pinning fashion.
Event 7, 8, 9:...Repeat....

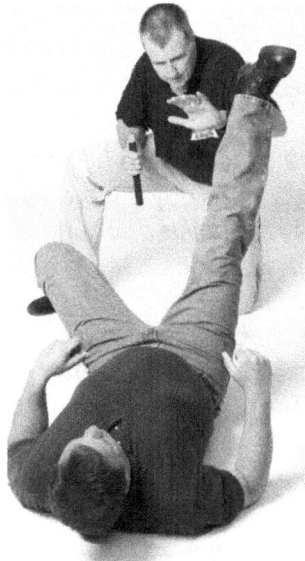  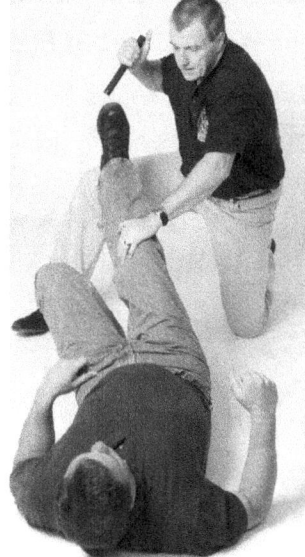

1: Trainee blocks.   *2: Trainee passes.*   *3: Trainee pins.*

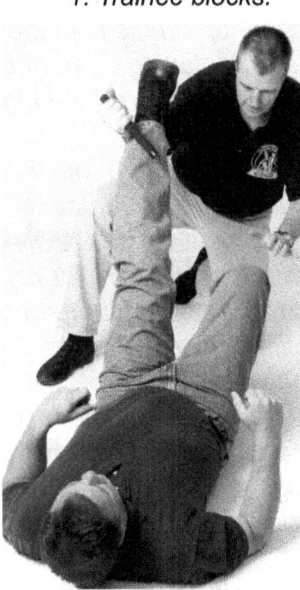  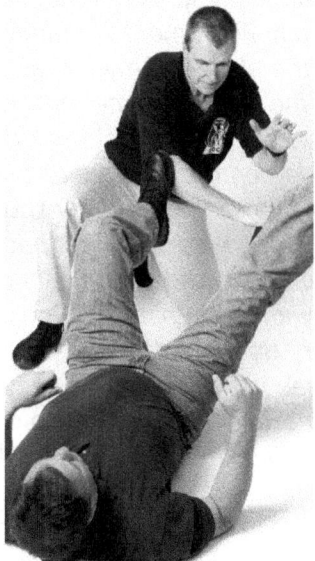

4: Trainer blocks.   *5: Trainer passes.*   *6: Trainer pins.*

1/2 Beat Inserts and Interruption Examples

* half-beat slashs to the lower leg, upper leg, and groin.

* half-beat stabs to the lower leg, upper leg, and groin.

* continue to invent half-beat leg options.

Black Box Knife Combat Files

The setting: Europe, Winter, World War II –

"We were securing the hillside. Bodies, our guys, their guys, were everywhere. They lay twisted around like they had no skeletons. Georgie and I, and some of the other guys were spread out checking the bodies. Turning the Nazi's over. Looking for weapons. Turning our guys over. Marking them for pickup with red strips of cloth on sticks. Georgie and I always poked the Nazis with our knives, you know to see if they were alive. I had my rifle over my shoulder. Sticking and turning bodies. Body after body. I guess I was daydreaming. I stuck another Nazi body, and the guy jumped right up on his knees and swung a knife at me! His knife was hidden under his chest. He was playing dead, the rat. He cut the sleeve of my jacket. I jumped back and fell on my back. He scared the high, holy, living shit out of me, now he was coming up at me trying to kill me.

He swung the knife at my legs, and I kicked at him. He mostly struck my boots, but he got my leg on the shin. I didn't feel it though. I stumbled, fell back and crawled back, kicking. My rifle slid right off my shoulder and, before I knew I was losing it, I was crawling back so fast, I crawled right back out of the sling. And he knew it too. He went for my rifle. I realized I still had my knife in my hand. I got up to a knee and, knife-first, I jumped on him. My knife got him good on his arm as he reached for my rifle, and we tumbled over. We were on our sides.

He had my arm with the knife. I had his arm with the knife. I tell you I saw the face of a ghost there looking at me. He was so white. Big, big eyes. I started kicking him, but he had a big jacket on, and I don't know if any caught him. Our legs pushed us apart. Trying to get up, he hit me in the side of the head with the knife and on the helmet with another hit. I saw Georgie behind him, trying to get his bayonet on his rifle.

I shouted, "Shoot him!" to him. "Shoot him!" But Georgie threw his rifle down and jumped right on the Kraut bastard. Right on his back and stabbed into his shoulder and neck with the bayonet in his hand. They rolled over, and I jumped on them both. The Nazi got up on one knee, I don't know how, holding Georgie with one hand and pushing me out. But we stabbed him. We stabbed him and stabbed him. Each stab you could tell he was getting weaker and weaker. He fell on his back. We fell on him stabbing. It was like we were punching him, but we had these knives in our hands. When we rolled off, he was dead. We just lay there in the snow, gasping for air on our backs.

"Why didn't you shoot him?" I asked Georgie. Georgie said that he was afraid to shoot because I was too close. And he couldn't get his bayonet fixed on, so he just held it in his hand."

– Alfred Bastacho, *Winter War, Wineheart*

CHAPTER 23: THE WINDMILL EXERCISES

This drill recognizes and develops fighting skills in the common downward reverse grip attacks. It contains two events as a basis and develops a multitude of tactical interruptions. The trainer acts like a moving training dummy. This builds coordination, speed and target acquisition to name a few. It is a classic Filipino martial arts drill.

Event 1: The trainer attacks from a high, over head stab. The trainee evades and passes.
Event 2: The trainee attacks from a high, over head stab. The trainer evades and passes.
Event 3: Repeat...

 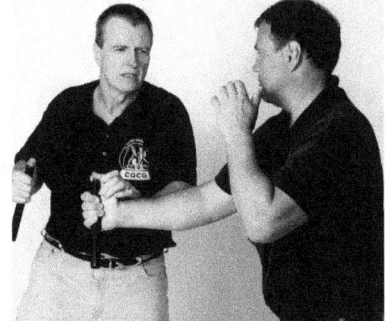 

The trainer attacks from a high, over head stab. The trainee evades and passes.

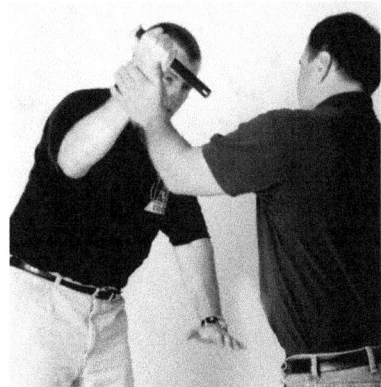 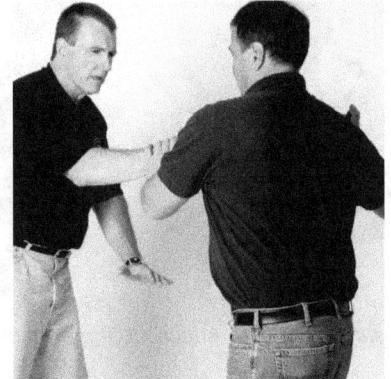 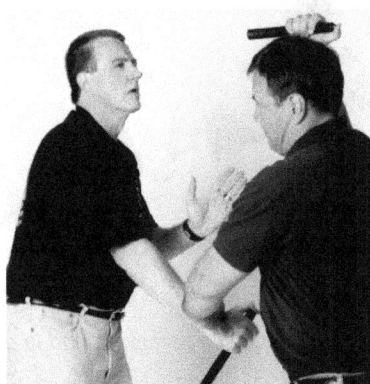

The trainee attacks from a high, over head stab. The trainer evades and passes.

From this windmill motion comes many practical moves.
    1: The trainer forearm blocks. The trainee forearm blocks and the two hit and retract until the trainer decides to change and return to circular pattern passing.

    2: The "small windmill." The trainer comes in at a 2 o'clock angle to knife tip cut the side of the trainee's neck. The trainee passes this and tries the same, until the trainer decides to change and return to circular pattern passing.

    3: The "chain saw." The trainer diverts from a big vertical circle to a small horizontal pattern to stab the stomach. The trainee copes this. This rotation continues until the trainer decides to change and return to circular pattern passing. For the basic chain saw, see the photos on the next page.

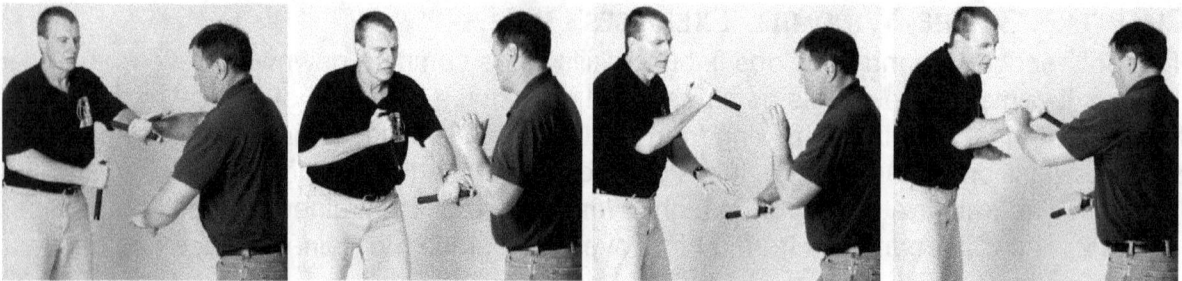
The trainer tries to stab the midsection. The trainee catches the arm, pushes down and tries to stab.

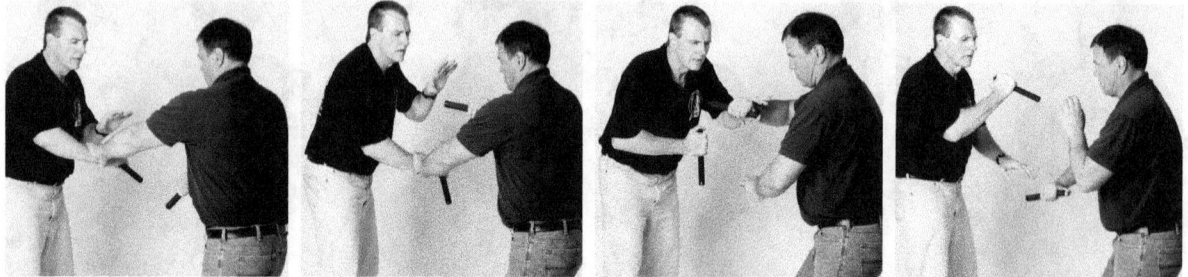
Then the trainee tries to stab the midsection. The trainer catches the arm, pushes down and tries to stab.

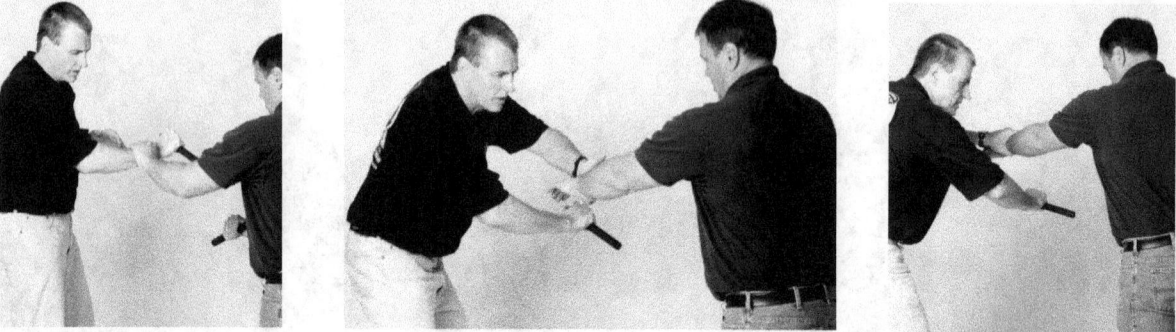
Trainee lifts the catching hand in a push/pull manner. Attacks to the armpit? Underneath the arms to the torso?

Windmill Chain Saw 1/2 Beat Inserts and Interruptions

    Option 1: Trainee pushes arm down, knife strikes the bend of the trainer's arm. The support hand grips the forearm and pushes it up so the trainer's arm is vertical. The support hand inserts into this bent opening as the knife extracts out in a slicing motion, getting a rear-arm bar, hammer lock. Any takedown or finish from there.

    Option 2: Trainee lifts the trainer's pushdown on the trainee's forearm. This frees the trainee's knife for any attack to the trainer's torso. Any takedown or finish from there.

    Option 3: The trainee tries to slice the trainer's forearm as he raises his knife. Continue to develop more responses out of this format.

There are many 1/2 beat tricks and options to the core, classic vertical Windmill, such as:
- grabs and yank backs.
- crossover moves.
- snake disarms.
- rear arm bar hammerlock captures.
- steerings to targets such as the thigh stab.
- more.

To show them all would involve making a whole book on this subject. Continue experimenting with the format and attend in-person sessions to pick up more.

Black Box Knife Combat Files

Ranger and Commando History...

" In the trenches, alleys, foxholes, and ruins, houses and bunkers of war, and in the drama of its espionage, knife combat has always existed. Whenever and wherever a soldier needed silence, or ran out of bullets, or couldn't reach his rifle or pistol (the majority are not issued pistols), or couldn't shoot due to proximity of comrades and explosives, the knife has left its scar or corpse ... as in this "non-medal" case of a joint British Commando/U.S. Ranger combat in France:

"After a few minutes of shelling, one of the mortar rounds found the main battery and there was a tremendous explosion. The dust and debris were still settling when the commandos charged in yelling and screaming. Koons and Sgt. Stempson watched awe as the British Commandos worked with bayonets and knives and cleared out the remaining enemy in the gun emplacements. While that was going on, other commandos were setting up charges to ensure total (and safe distance) destruction of the battery."

– Ian Padden,
U.S. Arborne/U.S. Rangers, Bantam Books

Black Box Knife Combat Files

Stabbed in the Head. "It felt like a sucker punch."

"He said, it felt like "a nasty sucker punch." Yet when he strained his eyes to the hard right, there was something that didn't belong: the pewter-colored contour of a knife handle of a 9" to 10" blade jutting from his skull. Sgt. Dan Powers, stabbed in the head by an insurgent on the streets of East Baghdad, triggered a modern miracle of military medicine, logistics, technology and air power. His survival relied on the Army's top vascular neuro surgeon guiding Iraq-based U.S. military physicians via laptop, the Air Force's third nonstop medical evacuation from Central Command to America, and the best physicians Bethesda National Naval Medical Center in Maryland could offer.

East Baghdad is a crumbling maze. Narrow lanes form stucco canyons that block out sunlight. A grimy film seems to blacken every surface: the facades, cobbled foot paths and street urchins' faces. Lines of sight end at each bend in the street, and the windows overhead look down like hundreds of eyes."It's just very slummy, with all these twisty alleyways," said Powers, now 39. "It's a nightmare to patrol."

A 12-year Army vet on his second deployment to East Baghdad, Powers spent his days training local police and trying to keep peace in a fortified cityscape. Soldiers in his 13-man squad would cruise the city's oldest quarter with Iraqi officers conducting street-level investigations and responding to gunfire or explosions. Nothing was different July 3 — at least not at first.

Powers was dispatched from Forward Operating Base Shield to a stretch of bomb-charred road. Explosive ordnance disposal personnel were already huddled over a blast site near Beirut Square on one of the district's wider thoroughfares. The explosion seemed minor so Powers and a team leader, Sgt. Michael Riley, were mostly concerned with warding off pedestrians. Powers was walking away from the cordoned area when it hit him — a near-knockout blow that felt like a 'clothesline tackle,' he said. But Powers stayed on his feet, spun around and slammed his raven-haired assailant to the asphalt, prodding the skinny Iraqi man's face with his M4 barrel. Riley, his squad mate, pounced and detained the assailant.

"I remember being pretty pissed off," Powers recalled to Air Force Times. Adrenaline throbbed in his veins and blood soaked his shoulder. A medic, Spc. Ryan Webb with the 118th Military Police, was tugging at his arm, demanding that he 'sit down, calm down and leave the knife in.' The knife? What knife?

"They said, 'You're stabbed' and ... I remember seeing the handle," Powers said. "There was no pain because the brain has no pain sensory nerves. It was all surface, like someone punched me in the head." Powers stayed conscious as soldiers carried him to a Humvee, sped to Forward Operating Base Shield and, after medics bandaged his head in clumps of cottony gauze, shuttled the sergeant to Baghdad's Green Zone.

Stabbings of American military personnel in Iraq or Afghanistan are extremely rare, outnumbered by drownings, strokes, cancer, drug overdoses and electrocutions. According to Defense Department casualty reports, Powers is only the second service member stabbed while supporting Operation Iraqi Freedom.

Doctors said, "Powers' injury had to be the most amazing thing anyone in the room had ever seen," Teff said. "An X-ray revealed that the knife entered just below Powers' helmet, above his cheekbone, skating right along the base of the cavity where the temporal lobe of your brain lives,"

Initially, they feared Powers, still in critical condition, could wake up with severe paralysis, brain damage and lost eyesight. But when the soldier surfaced after four comatose days, a battery of tests proved the stabbing had not robbed his intelligence or memory."

– Patrick Winn - Staff Writer, *Army Times*, Oct. 2007

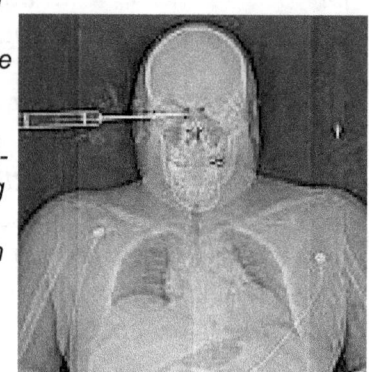

Army photos of the x-ray

Chapter 24: The Knife Horizontal Blast

In the martial arts world, the word "blast," or even the term, "horizontal blast," appears with some frequency. This is a knife version and runs horizontally and its maneuvers resemble the classic "blast" ideas. This process can be done outside the attacking arm or inside the attacking arm. For our example, we will use outside the "same-side' arms. Two training partners stand before each other with saber grips. The trainer stabs face high.

The duo stands ready for the drill.

Number 1 stabs at the face and neck.

Number 2 dodges and palm slaps the attack aside.

Number 2 hooks the knife up and pushs the attack.

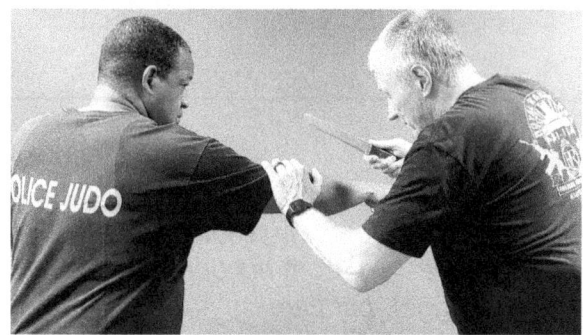
Number 2 further pushes with the hand again.

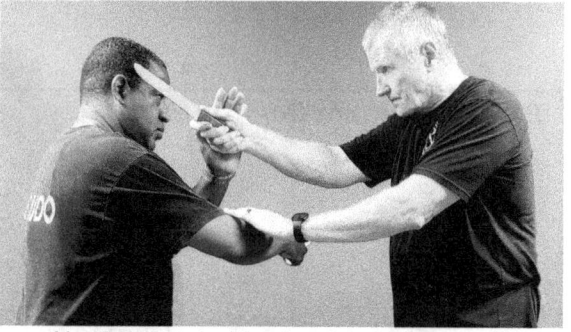
Number 2 stabs Number 1 face, neck high. Number 1 repeats the steps and....

...Number 2 stabs Number 2 again.

Step 1: Stab and stab blocked by hand.
Step 2: Knife swings in to also block.
Step 3: Hand block moves up to elbow and stabs back.

Steps 4, 5, 6: Repeat for both sides.

Most practitioners agree that this is a three-beat process for each side for a total of a 6 beat event. Once this is practiced you can start adding 1/2 beat attacks. A very common one is the 1 and 1/2 beat, hooking, uppercut stab into the jawbone.

There are numerous changes and variations one can do in classes and seminars. There is a two beat version, which resembles a Wing Ching Kung Fu drill, best passed on, felt and practiced in person, then shown in photos.

Black Box Knife Combat Files

"Gurung walked through the huts assured of their clearance, with his rifle slung snugly over his shoulder. ..he saw the Indian. Gurung drew the kukri from his belt. The soldier saw this and pulled his dagger, but Gurung had the start! Gurung swung at the man's arm as it came up with the knife, and he hacked into it below the elbow. The hand became lifeless. Gurung swung back and slashed open the throat with the curved tip, and the Indian fell. Gurung brought his rifle to bear upon the tribesman, but he was dead, the arm under the jacket, hanging on by muscle threads."

– Colonel Richard Mounds (Ret.),
My Military Life in Nepal

FORCE NECESSARY: KNIFE!
KNIFE FIGHTING

Knife vs. Hand - Knife vs. Stick - Knife vs. Knife - Knife vs. Gun

Knife Dueling

Chapter 25: Knife Dueling

Knife-to-knife dueling is a controversial subject. I have come to believe that knife dueling is way over-emphasized and over-practiced in these so-called "reality" knife training courses. This is something I have long called – "the myth of the duel." The "myth of the duel" is complex subject in the splitting and organizing of martial arts and survival training. (You don't learn how to play basketball to become a football player.)
Too many knife practitioners, fooled or ignorantly thinking they are studying realistic, modern or military knife combatives, express themselves through too much knife versus knife dueling. A methodology that is a mythology.

If you should escape a prisoner of war camp with a sharpened butter knife, the people who hunt you down have machine guns and dogs. It is unlikely you will be in a Rathbone-Fairbanks duel. Though it has happened in peculiar military circumstances as I have recorded in my Knife Combatives book. It took extensive searching into autobiographies, biographies and history books, here in the age of firearms, to collect military knife duel events. They are quite rare in the big picture of combat. There are a few more civilian-criminal events than military. The second murderer I caught in the act, in Texas, had killed a rival in a bloody. kitchen-knife duel!

We in modern times live in a hand, stick, knife and gun, mixed-weapon world and a stand-off duel of sorts is not common. Still we must practice a proportionate, appropriate amount of knife versus knife dueling because the uncommon event has and will occur. We always need many skills in combinations, slashing, stabbing, support strikes and kicks, footwork and many aspects of knife awareness.

For example, in the "who, what, where, when, how and why of life", if you are standing there with a knife in your hand, in front of another person with a knife? Why are you still there? If at all possible, an orderly retreat is in order. You better have a good reason to stay!

I think knife course instructors may knife spar at each and every one of their own classes and seminars for exercise as they wish, as long as they teach and grasp the Myth of the Duel concept. The legendary Dan Inosanto said once in a seminar I attended, "knife dueling is really about developing footwork." Instructors have different reasons for pursuing the subject. History? Fun? Competition?

Reality knife dueling can occur! They have happened. But common instructors usually forget the stress quick draw, the usually complicated, often strange, overall situations, and the physical layout of indoor and outdoor grounds/flooring where duels occur. These are overlooked factors in reality dueling training.

Strange places? I worked a murder case once where a big-knife, Bowie versus K-Bar, duel occurred between the driver and passenger in the cab of a big lumber truck, traveling down a two-lane highway! Driving and dueling. The driver won! When survival training we should work on the obvious things first, and not spend a lot of time on things less likely to occur. Once this doctrine has been proportioned, we can delve into the less likely, because, as I have said, these things happen too!

The same holds true for stick fighting. It is unlikely most people will be in a 28-inch stick fight, duel. Of course, if you do these things for fun, as a hobby? As a sport? Go for it! I am happy if you are happy. I just hope people know what they are doing, and why they are doing what they are doing in the big picture. (As I said earlier, you don't learn how to play basketball to become a football player.)

If and when two participants are actually forced to engage in a knife-versus-knife duel, the basic physical training of the battle is constructed as follows in a give-and-take series progression. These following dueling constructs are really a review of everything we have just covered in this book.

> Construct 1: Evade/invade: The principles of footwork. Footwork/environment.
> Construct 2: Strikes.
> Construct 3: Fakes.
> Construct 4: Blocks dodges/evasions.
> Construct 5: Counters to Common Blocks.
> Construct 6: Counters to the counters - strikes more strikes.
> Construct 7: Counters to the counter strikes - more blocks.

Construct 1: Evade/invade: The principles of footwork. Footwork/environment
Construct: There are three "participants" in a two-person, face-off duel. You. Him, and the "room," the arena, or the battleground. This is the "where" of the who, what, where, when, how and why survival questions. Your ability to stand ground, move or retreat is dependent upon the layout and conditions of your location. It is wise to exploit the layout and conditions. Forgetting the overall situation and the layout of grounds where a duel occurs is one of the most single overlooked factor in reality dueling training.

Construct 2: Strikes:
> *a:* The knife strikes four ways. The tip. The edge. The flat of the blade. The pommel. The knife slashes or stabs in a saber or reverse grip. The support, empty hand strikes with the hand forearm, elbow, shoulder. The support legs kick
> with the foot, shin, knee. The empty hand may grab.

Construct 3: Fakes. The science of fakes and feints. Feint is a French term that entered English language from the sport of fencing. Feints are maneuvers designed to distract or mislead, done by giving the impression that a certain maneuver will take place, while in fact another, or perhaps even none will. In military tactics there are two types of feints: feint attacks and feint retreats. All weapon and empty hand attacks invade in power lunges, pumps, hook angles or thrusting angles.

Construct 4: Blocks and dodges/evasions.
    a) The knife and weapon-bearing limb blocks incoming attacks.
    b) The empty or other limb blocks incoming attacks.
    c) The body evades with ducks, dodges often with minimum-to-no footwork.

Construct 5: Counters common blocks three ways.
    a) Cut or hurt the blocking limb.
    b) Re-direct your attack on another line of attack.
    c) Invading hands-use your hand and forearm to execute the four Ps.
    d) Pinning, Passing, Pulling or Pushing on the blocking limb to clear a path for
       a major or best target. Not the closest target, the best target.

Construct 6: Counters to the counter strike - more strikes.
    a) You counter the block with more strikes.
    b) Repeat the strikes, fakes and feints.
    c) Counters to the counter strikes - more blocks and dodges/evasions.

Construct 7: Counters to the counter strikes - more blocks.
    a) Repeat the above process

Dueling "stance/stand-off" knife and hand maneuvers

Duel Point 1: Knife is still and strikes suddenly.
Duel Point 2: Knife travels in a figure eight motion.
Duel Point 3: Knife goes up and down.
Duel Point 4: Knife goes in and out.
Duel Point 5: Knife goes side-to-side.
Duel Point 6: Dueler uses one, or some or all of the above.
Duel Point 7: Faking as expressed through the Basic 4 or Advanced 12 Combat Clock.
Duel Point 8: Lures. The knife retracts in a poor defensive position to set up a predictable counter-attack, such as:

    Lure 1: Open your belly, hit his incoming attack.
    Lure 2: Offer knife hand, hit his incoming attack.
    Lure 3: Offer free hand, hit his incoming attack.

Duel Point 9: The support hand works with these motions:
    a) holds another weapon as an:
        - impact tool.
        - shielding device.
        - projectile.

    b) opposite the knife movements.
    c) with the knife movements.
    d) in a guarded position independent of the movements.
    e) flicks and fans for distraction.

Duel Point 10: The Diminished Fighter Theory (DFT)
DFT is an aggregation of wounds and/or a loss of fitness performance.

Defense and Counter Attack Dueling
    Review the Dodge/Evasion Drill from prior segment.
    Review basic footwork moves and obstacle course applications from prior segment.
    Review knife blocking and deflecting from prior segment.
    Review support hand, checking, deflecting, passing from prior segments.
    Review and practice adding counter-attack options.

        1: eye jab.
        2: stab key or best targets.
        3: slash key or best target.
        4: grab and any attack.

Knife Battle Sets
In these duel training drills, two duelers start out facing each one. One dueller takes any number of steps forward as he strikes out with his knife, with each step. The other dueller steps back the same amount of times with blocks to these attacks. They repeat the moves, changing aggressor/defender roles. This is an early training method to establish. Start with 3 because it's slower and easier. As you reduce the sets, it becomes quicker and harder.

Battle sets of 3
In these duel training drills, two duelers start out facing each one. One dueller takes any 3 steps forward as he strikes out 3 times with his knife, with each step. The other dueller steps back 3 times with 3 blocks to the 3 attacks. They repeat the move, changing aggressor/defender roles.

Battle sets of 2
In these duel training drills, two duelers start out facing each one. One dueller takes any 2 steps forward as he strikes out 2 times with his knife, with each step. The other dueller steps back 2 times with 2 blocks to the 2 attacks. They repeat the move, changing aggressor/defender roles.

Battle sets of 1
In these duel training drills, two duelers start out facing each one. One dueller strikes out once with his knife. The other dueller blocks the attacks. They repeat the move, changing aggressor/defender roles.

Sets of 3 attack and defend practice.		Sets one-hand saber.
Sets of 2 attack and defend practice.	X	Sets one-hand reverse.
Sets of 1 attack and defend practice.		Sets two knives versus two knives.

Knife Dueling Invasion: The Invading Knife

In the duel, the duelers attempt to get close enough to score significant damage. Often, the entry is blocked by the enemy's limbs. This is an obstruction, usually knife arm to knife arm, or knife arm to empty limb side. The arms hit like reference points, either high or low. The reference points and entries are common and must be trained for problem-solving.

Common dueling entries to key targets are:
 1: entry over the arms.
 2: entry under the arms.

Common reference, contact points are: outside the arms.
 1: forearm to forearm.   via <
 2: forearm to upper arm.   inside the arms.

Invading Knife – The Basic Training Counters to Common Blocks, the 4 Ps,
Once the arm contact and obstruction is made, a dueller may clear the obstruction by pinning, passing, pulling or pushing on the opponent' limbs. These moves may be best implemented against a weakened or wounded enemy, or an untrained one.

Pinning the knife arm is pushing it up against the body, or pushing it against the walls or even the ground or floor.

Passing the knife arm.

Pulling the knife arm.

Pushing the knife arm.

Basic Training Knife Dueling Fakes

The science of faking is easily introduced by making applications to the Combat Clock. The soldier will strike at a certain angle, "sell it" really well with authentic motion and see that the enemy has reacted to it. If the enemy does react accordingly, as in moving to respond? The fake has worked and the soldier efficiently strikes on another opened line of attack.

A sample fake. A fake at 12 o'clock or the high line. Another line of attack is opened as the opponent moves to block the original high line attack. The defender re-directs his attack on another open entry line.

Basic Training Faking Work Out List:
This work list introduces the faking concept to the defender.

    Fake on 12 o'clock high:
        Option 1: strike on 3.
        Option 2: strike on 6.
        Option 3: strike on 9.
        Option 4: strike on 12 after a hesitation, that line might become open.
        Option 5: kick the leg.

    Fake on 3 o'clock, right side:
        Option 1: strike on 12.
        Option 2: strike on 9.
        Option 3: strike on 6.
        Option 4: strike on 3 after a hesitation, that line might become open.
        Option 5: kick the leg.

    Fake on 6 o'clock low:
        Option 1: strike on 9.
        Option 2: strike on 12.
        Option 3: strike on 3.
        Option 4: strike on 6 after a hesitation, that line might become open.

Fake on 9 o'clock, left side:
> Option 1: strike on 12.
> Option 2: strike on 3.
> Option 3: strike on 6.
> Option 4: strike on 9 after a hesitation, that line might become open.
> Option 5: kick the leg.

Advanced Training Faking Work Out List:
Rotate the angles though the 12 numbers of the Combat Clock.

Basic Training Dueling Events
The following examples are common dueling events to incorporate in dueling practice.

Over-Chambering and Over-Swinging Situations
In the quick, emotional violence of a knife encounter, the soldier might rear back too far in hopes of inflicting a deep wound deliver and a powerful slash. Should that slash miss, the soldier's weapon arm might travel too far. The slash could even cut the enemy, but still slide off and travel too far. Both of these over-movements, over-expose the defender to a counter attack.

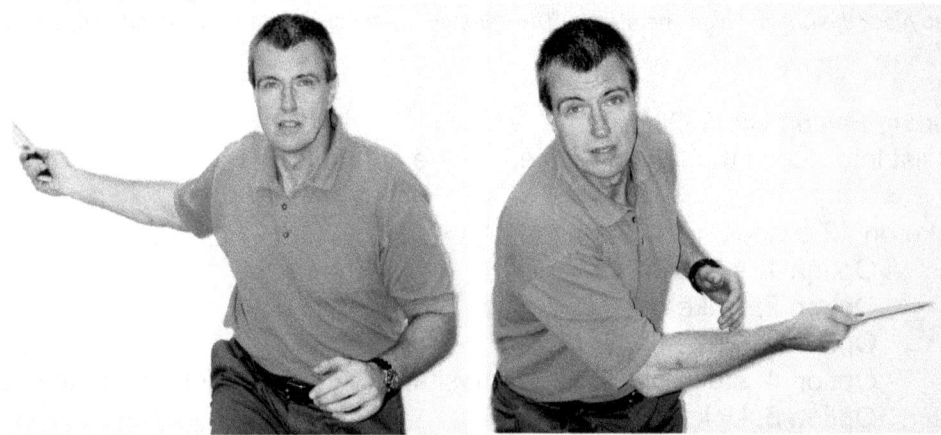

Reared back too far.   *Swung too far.*

The defender must keep all his or her slashes as efficient and as tight as possible, as previously written. Train this response in a training drill.

Over-Chamber/Over-Swing Duel Drill:
  Step 1: Trainer stands before trainee.
  Step 2: Trainer rears back suddenly, and the trainee stabs the trainer.
  Step 3: Trainer resumes a stance before the trainee.
  Step 4: Trainer slashes at trainee and over-extends his slash.
  Step 5: Trainee pins/controls the trainer's weapon arm and stabs the trainer.
  Step 6: Trainer resumes a stance before the trainee.
  Step 7: Trainer attacks in unpredictable free-style.
  Step 8: Trainer and trainee knife spar/duel. Trainer feeds these moves inside other dueling tactics.

Mike Gillette demonstrates an over-swing, exposing a center-line weakness in his defense.

Gillette demonstrates a power slash, miss and an over-swing, exposing his side to a pin and stab.

Knife forward versus knife back situation
Many militaries suggest a knife-back stance when fighting knife versus knife. If both duelists are knife-back it equalizes the opponents. If one fights knife-forward, he will usually have a distinct advantage versus the common knife back dueller. The knife forward duelist should work to destroy the opponent's lead limb.

In and out situations
One of the common occurrences in all knife dueling is moving in to strike, then moving right back out to evade a counter-attack. This can be trained in the following drill manner. The trainer holds up a stick to be struck by the trainee. While the trainee moves in to strike, the trainer swings another padded weapon at the trainee, causing him to be wary of the counter-attack and to learn to evade the common counter-strike. The trainee may also learn to manipulate his primary strike into a quick, follow-up secondary block.

Here Gillette demonstrates the knife-back stance, thinking his lead arm will "cover and protect" his knife. Attack the lead arm. The results to the lead arm might be minor, moderate, severe wounding or even maiming should knives be big and sharp enough.

In and out situations

One of the common occurrences in all knife dueling is moving in to strike, then moving right back out to evade a counter-attack. This can be trained in the following drill manner. The trainer holds up a stick to be struck by the trainee. While the trainee moves in to strike, the trainer swings another padded weapon at the trainee, causing him to be wary of the counter-attack and to learn to evade the common counter-strike. The trainee may also learn to manipulate his primary strike into a quick, follow-up secondary block.

The trainer, armed with a wooden stick and padded knife, develops the in and out footwork and punishes the trainee for not escaping back fast enough after coming in to attack.

Knife Dueling/Sparring Exercise List

The many combinations duelers can practice:

Right hand vs. right hand grip.
Left hand vs. left hand grip.
Right hand vs. left hand grip.   X
Left hand vs. right hand grip.
Saber grip.
Reverse grip.

1 knife vs. 1 knife.
1 knife vs. 2 knives.
2 knives vs. 1 knife.   X
2 knives vs. 2 knives.

1 person vs. 1 person.
1 person vs. 2 people.
2 people vs 2 people.

Knife versus knife dueling tactics could fill an entire manual alone. Dueling mastery is not the subject of this manual. Once a duelist enters into a closer combat range, hands-on and grappling strategies begin, which are tactics not technically classified and taught as dueling by many training systems. All these other tactics are detailed in the subsequent segments of this book. Knife-versus-knife dueling is only a part of military knife fighting, perhaps even a small, unlikely part in the modern, mixed-weapon world of combatives.

What is the Enemy's Other Hand Doing?

When working on knife combat scenarios, the soldier must access what the other soldier's hand is doing during the steps of a fight. The other hand may be up in a combat ready position at the face or shoulder, mid-height at about the rib cage, low or swung way behind him. The other hand could also be attacking in a thrust or hook attack. It could be empty handed or holding a second weapon, or shield of some sort.

The enemy's support hand positions

Enemy Hand Position 1: At or near the face?

Enemy Hand Position 2: On the chest?

Enemy Hand Position 3: At the lower ribs?

Enemy Hand Position 4: Hanging straight down?

Enemy Hand Position 5: Swung back like a fencer?

Enemy Hand Position 6: Holding a second weapon of some sort?

Enemy Hand Position 7: Holding a shield of some sort?

Enemy Hand Position 8: Attacking in a thrusting line?

Enemy Hand Position 9: Attacking in a hooking line?

Enemy Hand Position 10: Moving about for distraction and deception?

A properly trained practitioner knows to monitor and assess the threat level of this other hand during the fight. The soldier should rehearse contingencies. While expecting the worst, many untrained enemies will not know to apply their second hand, over-concentrating and over-depending upon their knife or other weapons.

Killshot Knife Fighting Competitions

Many soldiers knife spar for training. Many carry the experience further by creating knife dueling tournaments. No tournament can replicate reality and all judging will be problematic. The challenge is properly identifying if the strike was or was not a kill shot and which ones were moderate to severe woundings. That is the fickle call of the referees/coaches supervising the fight. The calls will not be perfect. No tournament is. The typical Killshot Knife fight consists of two fighters and one or more referees and coaches. The referees should have a whistle or such loud means to interrupt a fight to declare a killing or a wounding. Fighters should wear proper safety gear, certainly a helmet and eye protection. The overall experience offers an abstract insight to a knife duel and develops athletic ability.

The Killshot Knife versus Knife Competition Rules

Rule 1: Duelers may fight with soft knives, wooden knives or metal training knifes. (Fools train with real knives.) Fighters agree to the knife. If one fighter selects a metal trainer and the other fighter selects a soft trainer, the fighters defer to the soft trainer.

Rule 2: The fight starts out as a duel and is primarily a duel, however the fights may turn into close quarter grappling and ground fights.

Rule 3: The Killshot victory is winning two-out-of-three round. Locals reserve the right to change that to 3 out of 5 fight, or whatever combination suits their goals.

Rule 4: Should a judge and/or referee, or a preponderance of judges and referees, agree that a seriously debilitating killshot to face, head, neck, heart, diaphragm, groin has been delivered, the whistle is blown. The fight stops. The deliverer wins that round.

Rule 5: Judges/Referees should take care to see if a mutual killshot strike was delivered simultaneously, or returned within a few seconds of the first strike, thereby rendering the event a dual death and dual loss with no clear winner. Both parties are credited with a loss, or a local rule may require a do-over – which does not support reality knife fighting. This rule is implemented to circumvent the common fencing rules. In fencing the first striker wins even if a half a second later the loser strikes back. Fencing has the doctrine, muscle memory mistake of failing to instill escape after a killshot blow has been dealt. Killshot fighting demands a safe escape from an immediate counter-attack.

Rule 6: Judges/Referees will determine woundings. Should a weapon-bearing limb be struck, the fight will be stopped and the victim is forced to change weapon hands. The local rules may implement punishment for the wounded limb. Two punishment possibilities are:

    One Wounded Limb Option 1: All five fingers are duct taped together to simulate wounding.
    One Wounded Limb Option 2: Wrist weight is strapped on the wounded wrist.
    One Wounded Limb Option 3: Both. Fingers are wrapped. Wrist weight added.

    Should this wounded person be significantly struck on the second limb:
        Two Wounded Limbs Option 1: Fingers on hands are duct-taped to simulate wounding.
        Two Wounded Limbs Option 2: Wrist weights are strapped to both wrists.
        Two Wounded Limbs Option 3: Both. Fingers are wrapped. Wrist weights added.

The fight continues on knife-versus-unarmed attacker. **Note:** many times the unarmed person still wins. The judges/referees will then have to decide what armed strikes, kicks and takedowns are winners.

Rule 7: Local competitions reserve the right to change up the encounters to create more challenging rounds.

Block Box Knife Combat Files

"U.S. Army Lt. R. Nett led his company into heavy Japanese resistance (at Ormac Valley, Guam). His troops were very low on ammo and heavily engaged in hand-to-hand, bayonet and knife combat. Lt. Nett low-crawled up to, and jumped into a foxhole, his knife slashing. There, he killed the first line enemy rifleman that was pinning down his men, and he continued onward, though seriously wounded, furthering his company's position. Taking shrapnel, and parrying off Japanese sabers, bayonets and knives, Lt. Nett used his knife to kill seven more enemy soldiers."

        – U.S. Army Synopsis Report,
        Public Affairs Office,
        Medal Winners WW II

Alternate Dueling

We have spent time and effort to experiment with knife versus knife dueling. Part of the knife defender experience is to work on:
- knife versus hand.
- knife versus stick.
- knife versus empty long guns.

Note: *As you obviously know, there is little, if any knife versus loaded pistols.*

Lots of practitioners work on being unarmed and facing knife attacks.

Lots of practitioners work on facing a knife attacker "while holding" an impact weapon.

Knife versus Unarmed

The old laws of fair play say the good guy doesn't fight an unarmed guy while holding a knife. Even the law can be mixed up on this. This "armchair" opinion dissolves when the opinionaire is off his chair and out on the dark side of life and suddenly a victim of a common crime of violence. When it's YOU versus him, maybe even a bigger, meaner, younger him? There have been annual USA statistics that say the presentation of a knife or gun scares off a criminal 65% to 69% of the time. Still, the knife is not God's gift to safety. The attacker may challenge the knife threat and it may not be as easy as one assumes to always win out. Experiment with this situation via dueling.

Knife versus Stick

Fighting an impact weapon should justify pulling a knife. The stick attacker can reach out and touch you a lot easier and the footwork and range problem is worse than knife versus knife, or hand versus knife. Again, the knife is not God's gift to safety. The attacker may challenge the knife with a "stick" and it may not be as easy as one assumes to always win out. Experiment with this situation via dueling.

FORCE NECESSARY: KNIFE!
KNIFE FIGHTING

Knife vs. Hand - Knife vs. Stick - Knife vs. Knife - Knife vs. Gun

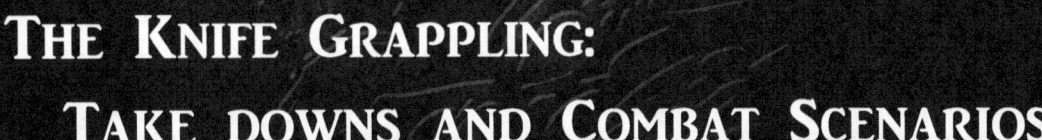

The Knife Grappling:
Take downs and Combat Scenarios

Chapter 26: Knife Grappling/Take downs

These diverse combat scenarios are just crisis rehearsals for defenders to spend time fighting with a knife in their hands, for all the obvious reasons. We start with a review of knife takedowns.

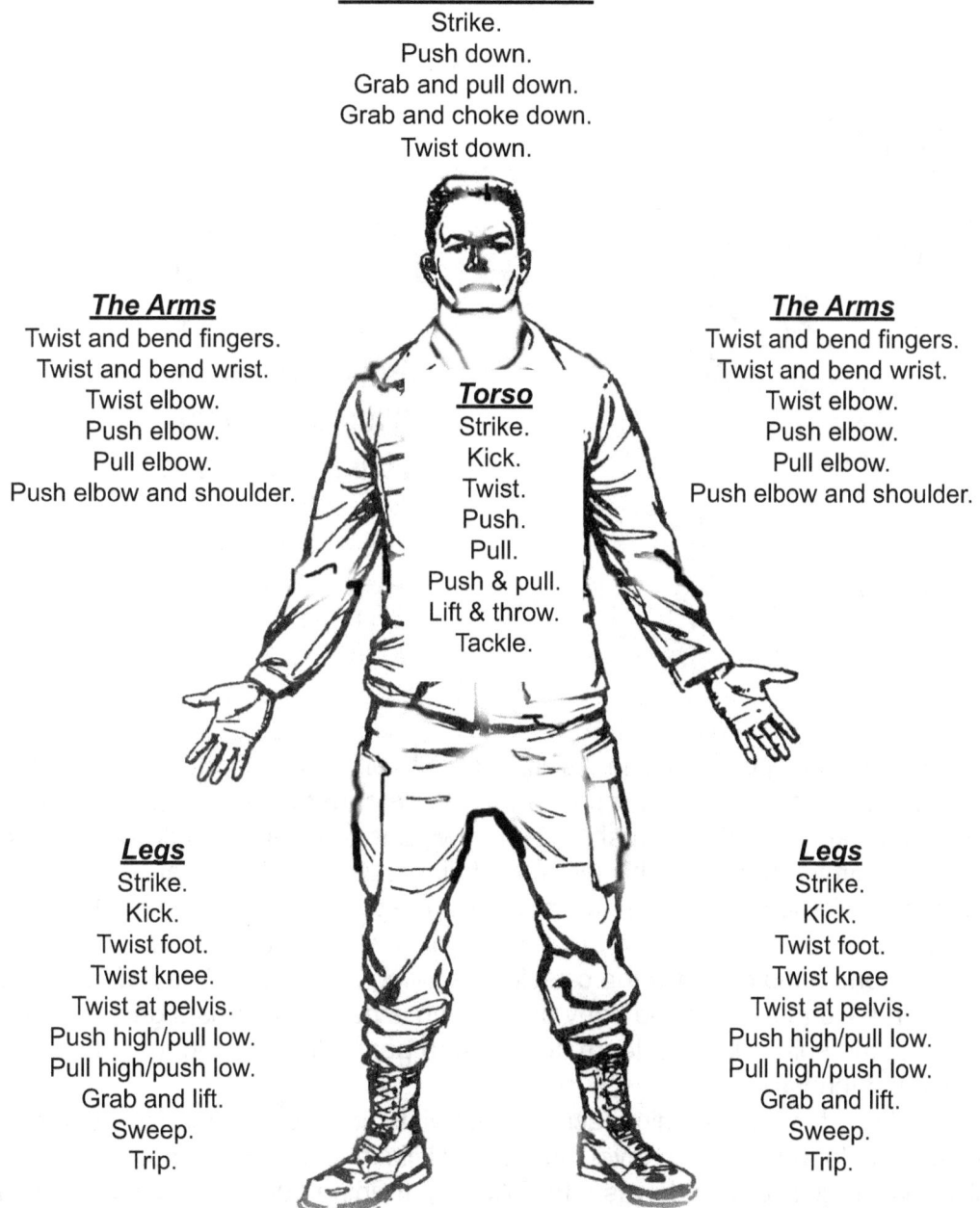

The Head and Neck
Strike.
Push down.
Grab and pull down.
Grab and choke down.
Twist down.

The Arms
Twist and bend fingers.
Twist and bend wrist.
Twist elbow.
Push elbow.
Pull elbow.
Push elbow and shoulder.

Torso
Strike.
Kick.
Twist.
Push.
Pull.
Push & pull.
Lift & throw.
Tackle.

The Arms
Twist and bend fingers.
Twist and bend wrist.
Twist elbow.
Push elbow.
Pull elbow.
Push elbow and shoulder.

Legs
Strike.
Kick.
Twist foot.
Twist knee.
Twist at pelvis.
Push high/pull low.
Pull high/push low.
Grab and lift.
Sweep.
Trip.

Legs
Strike.
Kick.
Twist foot.
Twist knee
Twist at pelvis.
Push high/pull low.
Pull high/push low.
Grab and lift.
Sweep.
Trip.

Even the most simple takedown almost always involves applications to multiple parts of the body. The pushing or pulling knife can cause body movements and takedowns not found in traditional, unarmed takedown training. In military knife combatives, the soldier must practice these applications while holding a knife in a certain grip. At times, some moves require certain grips. Some takedowns can be worked with a ice pick grip/reverse grip and not with a saber grip, and vice-versa. These applications can be learned through controlled experimentation and repetition training.

CUT 'EM DOWN,
KNOCK 'EM DOWN,
PUSH 'EM DOWN,
PULL 'EM DOWN,
TRIP 'EM DOWN,
TWIST 'EM DOWN.

Grappling is an important combat skill. A knife in one hand changes the unarmed combatives paradigm considerably. Empty-handed, a defender takes the enemy down to the ground by knocking him down, pushing him down and/or pulling him down, twisting his head, body or limbs, or tripping his legs. Most times, two free hands are an advantage in grappling and takedowns, and a necessity while executing more than half the official list of universal takedowns. But taking the enemy down while holding a knife in a saber or reverse grip in one hand changes the classic, grappling battle plan. The defender loses one gripping hand. He must hook the body part, which is not as solid as a grip. He must push and pull with the knife. He must learn not to use a percentage of common, two handed takedown and throws.

Given training time and training facility constraints, it becomes impossible to train large military units in long-term, high-level, grappling skills in an effort to create a virtual black belt skill level in "jujitsu" takedowns. The following training outline is a simple, scientific breakdown of basic grappling takedowns while holding a knife.

In order to execute takedowns while holding a knife, the soldier needs to practice body rams, empty hand striking skills, knife striking skills, support and weapon-bearing limb striking skills, grabbing and wrapping skills, and off-balancing and tripping skills. Rather than face the challenge of describing every takedown while holding a knife, which would be the subject matter of an entire thick book by itself, the following is an overall synopsis.

This synopsis added with the upcoming combat scenario chapters will provide a defender with a healthy working knowledge of knife takedowns.

Knife Takedown Study 1: "Cut 'em down"
Cutting the enemy down means using the knife to disable, wound and maim the enemy until he falls. This is accomplished by cutting the muscles or causing enough damage as to create significant blood loss.

Slash, hack and cut:
- Bloodlines.
- Muscles.

Knife Takedown Study 2: "Knock 'em down!"
The defender needs striking skills to stun, move and/or render the enemy unconscious, wounded or dead on his feet for takedowns. These skills would fill and entire training manual alone. They are only summarized here:

Support limb strikes:
- Finger strikes (usually to the eyes).
- Palm strikes.
- Hammer fists.
- Punches.
- Forearms.
- Elbows.
- Upper arm and body rams.
- Limited, careful use of the head butt.

Knife strikes:
- Saber knife strikes.
- Reverse grip knife strikes.
- Pommel strikes.
- Flat-of-the blade strikes.

Kicks:
- Stomp kicks.
- Inward round kicks.
- Outward round kicks.
- Thrust kicks.
- Back kicks.
- Side kicks.
- Knees.

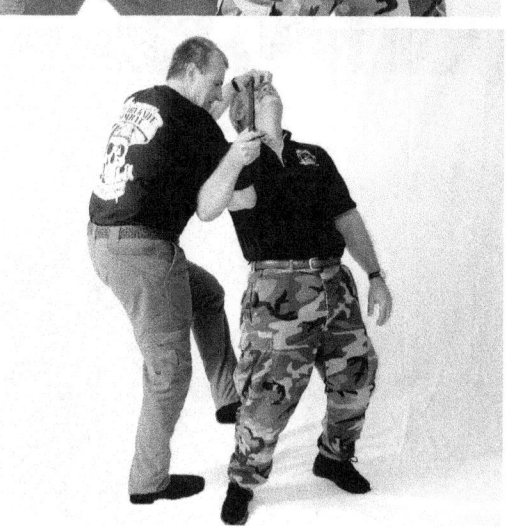

Knife Takedown Study 3 and 4: "Push 'em and pull 'em down'"

The defender needs pushing and/or pulling skills to take the enemy down. This may or may not involve grabbing and controlling some part of the enemy's body. A grab is executed with a hand, and arm or a hooking capture with a saber or reverse pick grip knife, as demonstrated in a prior section of this manual.

The defender grabs and/or catches with:
   Grab/catch 1: His support/empty hand.
   Grab/catch 2: The bend and hook of his knife and hand.
   Grab/catch 3: His support/empty arm wraps.
   Grab/catch 4: His weapon-bearing limb arm wraps.

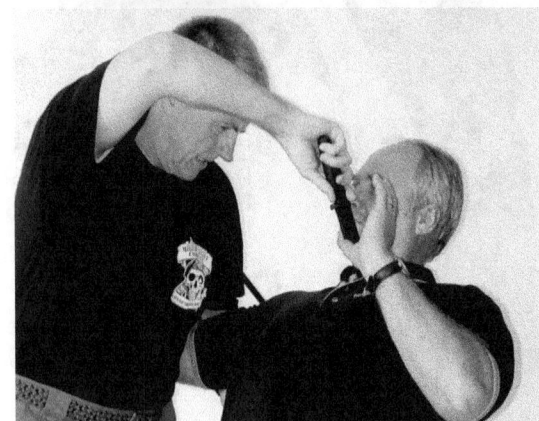

The defender:
   Pushes the enemy off-balance.
   Pulls the enemy off-balance.
   Pushes the enemy high and pulls low.
   Pulls the enemy high and pushes low.

The defender can push or pull with his knife
– with a saber grip.
– with a reverse grip.
– with the edge.
– with the tip.
– with the flat-of-the-blade.
– with the pommel.
– with the "hook" position between his
– hand and knife.
– with the knife-hand thumb raised to catch the enemy's wrist.

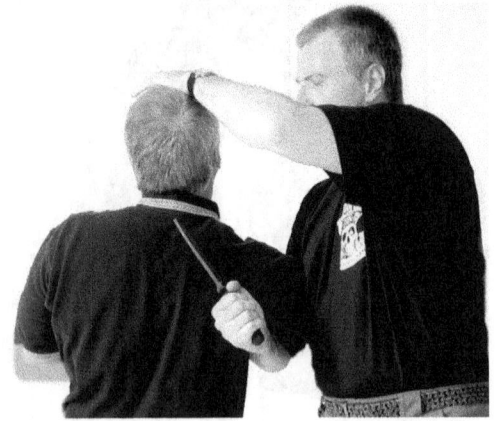

Knife Takedown Study 5: "Trip 'em down"

The soldier trips the enemy down by sweep, hooking his legs. Pull a captured arm if possible. Push against the upper torso. This should be executed with a simultaneous wheeling motion. To gain the advantage, step past the enemy, pulling him off balance in the direction of tripped leg. Militaries teach single and double leg sweeping trips and back/heel kicking the calf muscle. The following is a fundamental leg trip, sweep takedown knowledge. A sample of the many...

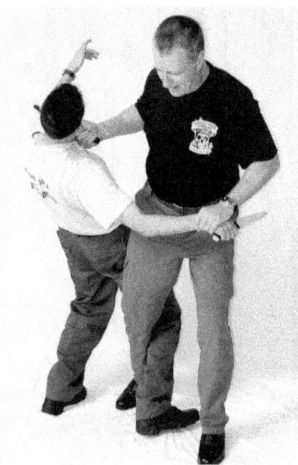 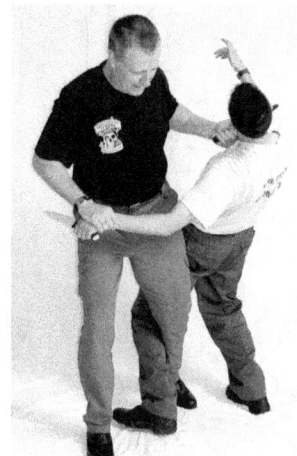

Outer right leg sweeping trip. *Outer left leg sweeping trip.*

Outer Leg Trip 1: Right leg sweep/trip to outer right leg with an upper body push, and arm pull.

Outer Leg Trip 2: Left leg sweep/trip to the outer left leg with an upper body push, and arm pull.

Outer Leg Trip 3: Right leg sweep/trip to both legs with an upper body push, and arm pull. Are his legs close enough together?

Outer Leg Trip 4: Left leg sweep/trip to both legs with an upper body push, and arm pull. Are his legs close together?

Outer Leg Trip 5: Heel kick to calf muscle area with an upper body push, and arm pull.

Inner Leg Reaps: If the height is shared and you can get a leg inside and get a hooking backward style "kick," with an upper body push, you can do these. Experiment as they are not easy.

Sometimes both legs are reach-able. Not always. *Back kick to calf muscle with either right or left leg.*

Right-side Inner leg reap trip. *Left-side Inner leg reap trip.*

Knife Takedown Study 6: "Twist 'em down!"

Twisting the enemy down means twisting the enemy's wrists, or arms, or head, or torso, or legs until he loses balance and falls down.

Considering that a defender may be fighting from downed, kneeling and standing positions, he can twist:

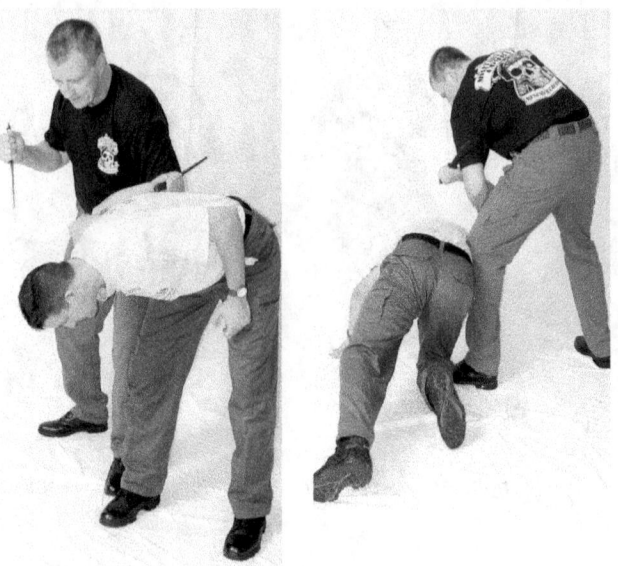

Sample of a bent-arm bar hammerlock twist takedown.

    The head and neck.
    The torso.
    The bent arm.
    The straight arm.
    The wrist.
    The finger.
    The leg.
        – ankle twist.
        – knee twist.
        – pelvis twist.

Summary: Tools of the Defender Knife Grappler

1: Striking skills to stun, move and/or render the enemy off-balance, unconscious, wounded or dead.

2: Disarming skills so that the defender can stun, disarm and move in more safely for a takedown. The three main disarms are the snake, the impact disarms and the push-pull.

3: Kicking to stun and imbalance the enemy, usually to the legs.

4: Grabbing skills and arm wrapping skills for pushing and pulling.

5: Support-side limb pushing, pulling and twisting skills.

6: Weapon-bearing limb pushing/pulling skills with saber and ice pick grip. This is done by manipulating the wrist into a hooking, catching and pinching position.

7: Leg tripping and sweeping skills.

CHAPTER 27: COUNTERING THE WEAPON THREAT PRESENTATION

This is a "when" question. When do you counter-attack into a possible grappling, combat scenario situation, dealing with the weapon threat stand-off. The enemy has a weapon out and is threatening to capture or control the defender and/or others. The defender finds the proper second, strikes the limb, observes a successful disarm, and finishes the fight as needed based on the mission with threats, or more knife attacks and/or a takedown.

When to strike. The enemy may be preoccupied with events in his environment and his eyes and attention may become distracted. This is especially possible with multiple people present and in a hostile, confusing situation. The soldier must watch the eyes of the enemy and await distractions if possible. This can be replicated in a training session by springing into action when a trainer cuts his eye to the side, or moves his head and eyes to look at something nearby.

Military, police and civilians have a tendency to orchestrate the personnel in their scenes with their weapons as pointers. A soldier must be alert to this propensity and watch for seconds when the enemy removes his weapon from bearing down on the defender to use it to direct and order, either his personnel or other prisoners around. The defender will have to draw his or her weapon and take action.

At times the enemy is orchestrating an attack, a capture or even a crime scene and becomes distracted by surrounding threats. The defender should look for openings when:
1: His eyes cut to the side.
2: His eyes and head cut to the side.
3: Head and weapon cut to the side in an attempt to threaten or control others in the scene, or to point a soldier where to walk.

Eyes cut to the side off of you.

Head and eyes cut to the side, off of you.

Head, eyes and threat arm moves to the left or right, off of you, in an effort to point, or threaten and/or control someone else at the scene of the incident.

Black Box Knife Combat Files

"I now had only one weapon with me, my Special Forces knife."
– Vietnam, Sgt. Roy P. Benavidez, USA SF (Ret.)

"Branches, slivers of wood, metal, dirt, and body parts were stinging us from the percussion caused by the bombing. We could feel the tremendous heat of the afterburners of the F-100s. That's how low they were flying. Gunships were diving and diving between the passes of the jets. The air support was like a swarm of killer bees attacking us. It later reminded me of that passage of scripture from the Book of Revelations about the sky turning black with locusts. Through the middle of this moment of hell came a lone slick that touched down about 20 to 30 meters away. We knew that this was our last hope to leave alive. We loaded the last of our ammunition. This was it. Now or never. I learned later that the fighters had run their fuel down to a level that would not allow them to return to Phan Rang, so they diverted to Bien Hoa near Saigon, where they refueled and flew back to Phan Rang. I got to O'Connor and gave him his third shot of morphine. I also took another shot in the leg. We were under heavy fire again, and I wasn't sure what was going to happen to us, even though I tried to reassure O'Connor. He must've thought I was losing it because I don't think any of us really thought we were going to get out of there.

We were surrounded. The air attack managed to stop the assault for a few moments, but it was long enough for that single chopper to lower right in front of us, and a Special Forces medic, Sergeant Sammons, ran to us from the aircraft. Roger Waggie and his newly formed crew of volunteers, W.O. Bill Darling as crew chief and W.O. Smith as door gunner, came to our rescue. What I saw was the American fighting man at his best.

The two of us carried or dragged as many of the men as we could. But the NVA were firing directly at the chopper, shooting the men as they were lifted aboard. Two of the men were shot in the back as they tried to crawl to safety inside the chopper. I could barely see through the matted blood in my eyes due to shrapnel wounds on my face and head. Waggie's chopper was badly shot up. He and his co-pilot were shooting through their front windshield with their 38 pistols, while Darling and the door gunner and Smith were firing the M-60s at separate groups of NVA charging from the sides. Darling and Smith had volunteered to man a gun because they knew we were running out of men, and as officers they didn't have to volunteer for this situation. All I know is that because they did, soldiers would live.

I made another trip to find Mousseau. He was lying in the grass. I tried to carry him to the chopper. I didn't even notice when one of the NVA soldiers, lying on the ground, got to his feet. I also didn't notice when he slammed his rifle butt into the back of my head. I turned to look at him. Both of us were surprised, I because I hadn't seen him and he because I had turned around after he had delivered the blow, but he reacted quickly and hit me again. I fell, my head swimming in pain.

I now had only one weapon with me, my Special Forces knife. I reached for it, and when I did he pointed his bayonet at the front of my belly. Fortunately, he hesitated, and it gave me enough time to get to my feet. He sliced my left arm with the bayonet, and I shouted to O'Connor to shoot him. But he was too drugged to move, so I did the only thing I could. I stabbed him with every bit of strength I had left, and when he died, I left my ESP. knife in him. The last round in my stomach had exposed my intestines and I was trying to hold them in with my hands. I could see Mousseau lying on the floor, staring at me with his one good eye. I reached down and clasped his hand and prayed that he would make it until we reached Saigon, where the medics could help him.

Sadly, he would be among the approximately 200 men who died on both sides during that battle. I hoped that LeRoy was with us, that at east his body was going home to his family. I had loaded some bodies on the chopper."

CHAPTER 28: LESS-THAN-LETHAL KNIFE COMBAT TACTICS

At times, missions, rules of engagement and use of force standards require the capture, containment and control - not the death of an enemy. This is often called by professionals as "non-lethal measures," but military and law enforcement specialists recognize the evolving legal and perception facts that the term "less-than-lethal" is a smarter, more comprehensive phrase. Tactics and equipment designed not to kill and called non-lethal, might still actually kill despite the design and name, rendering the term non-lethal into an operational misnomer and confusing liability.

Words of surrender.

A comprehensive knife/counter-knife program also covers less-than lethal applications of the possible knife confrontation. The following tactics are less-than-lethal and can be substituted for lethal movements.

Less-Than-Lethal 1: Verbal skills and the art of surrender
Your weapon presentation, your threats, your disarm, or your initial display of skill in the onset of a fight may cause the enemy to surrender. At times, winning the ground and getting the tip of your knife against the enemy, along with a verbal threat, may coerce him to surrender.

The pommel may strike.

Less-Than-Lethal 2: The knife pommel strike
The pommel strikes is another less-than-lethal strike unless it cracks the skull.

Less-Than-Lethal 3: All support hand strikes and kicks
Support hand striking, and kicking the enemy are less-than-lethal moves. As the picture to the right demonstrates, the enemy has dropped his weapon and is theoretically an unarmed man and in many situations, both military and civilian cannot be killed. The knife hand punches. The defender may punch the enemy with the flat of his fist, forgoing the stab or slash.

The empty support hand may strike.

Less-Than-Lethal 4: The closed folder
The defender may close his or her tactical folder and use the folder as a "palm stick" impact weapon."

The "fist" of the knife grip may punch. Saber or reverse.

Less-Than-Lethal 5: Knife slashes on secondary targets
Slashing muscles and other secondary targets may cause an enemy to surrender or collapse, without a fatality.

Less-Than-Lethal 6: The flat of the blade strikes a stunning blow or can be used to push or pull in grappling. Many militaries teach the flat of the blade strike to the head of an enemy to stun and bewilder them, as a set-up for further action. When a less-than-lethal mission becomes mandatory, this flat strike becomes an option.

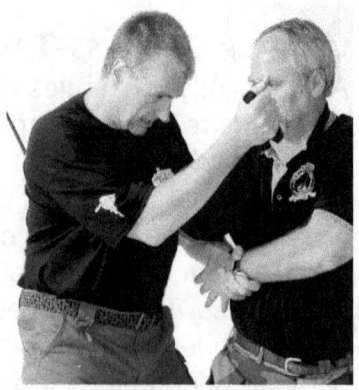

The closed folder may strike.

Less-than-lethal slashes may disarm or only wound.

The flat of the blade may strike, push or pull.

Black Box Knife Combat Files

France, World War 1
"We got the order to storm a French position, strongly held by the enemy, and during the ensuing melee a French corporal suddenly stood before me, both our bayonets at the ready, he to kill me, I to kill him. Saber duels in Freiburg had taught me to be quicker than he and pushing his weapon aside I stabbed him through the chest. He dropped his rifle and fell, and the blood shot out of his mouth. I stood over him for a few seconds and then I gave him the coup de grace. After we had taken the enemy position, I felt giddy, my knees shook, and I was actually sick."

– Richard Holmes, *Acts of War*

CHAPTER 29: BASICS IN STAND-UP, KNIFE COMBAT SCENARIOS

We will now look at using the quicker kill targeting within very basic set-up, standing combat scenarios. Each one starts with a strike and block and finishes with a take down.

Sample 1: Saber Grip Attacking the Brain (through the eyes and nose)

The enemy strikes.

The defender blocks and grabs the attacking limb.

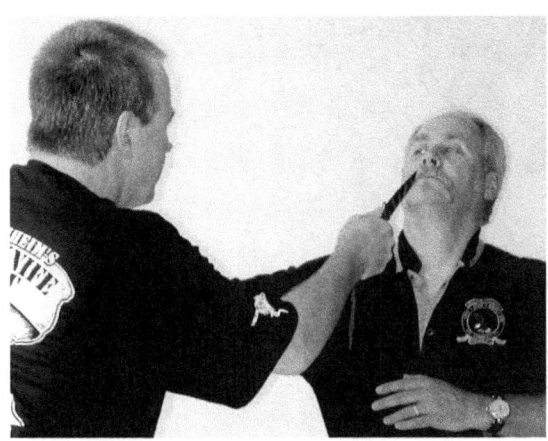

The defender drives his saber grip knife upward into the eyes or the nose, with the eventual target being the brain.

Any take down that suits your position.

2- Person Drill Practice:
    Step 1: The enemy attacks with any weapon, on the Combat Clock Basic 4 or Advanced 12.

    Step 2: The defender blocks and/or grabs the attacking limb and power stabs the center of the face.

Sample 2: Reverse Grip Attacking the Brain (through the eyes and nose)

The enemy strikes. The defender blocks and grabs the attacking limb.

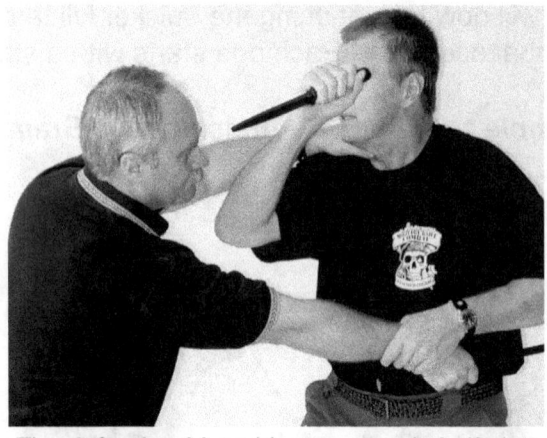

The defender drives his reverse grip knife into the eyes or up the nose, with the eventual target being the brain.

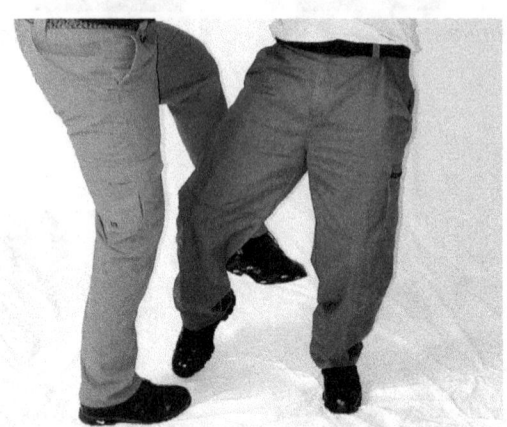

Any take down that suits your position.

2- Person Drill Practice:
    Step 1: The enemy attacks with any weapon, on the Combat Clock Basic 4 or Advanced 12.

    Step 2: The defender blocks and/or grabs the attacking limb and power stabs the center of the face.

Sample 3: Saber Stab to the Base of the Skull

Next in the progression is an isolated study in attacking the brain function through the base of the skull. The knife attacks a vital nerve center in the base of the skull hoping to disrupt the cerebellum, and/or the medulla oblongata and/or perhaps even the top of the spinal cord. The knife should be pumped or twisted upon entry.

I have seen this scenario in the Filipino martial arts, Indonesian Silat and US Navy World War II combatives. It starts with both opponents trying to execute stomach stabs, both slapped aside and the charges lead to linking arms. He or she that takes this action up after the link, is the survivor.

The enemy stalks and attacks.

The defender dodges and slap-passes the attack.

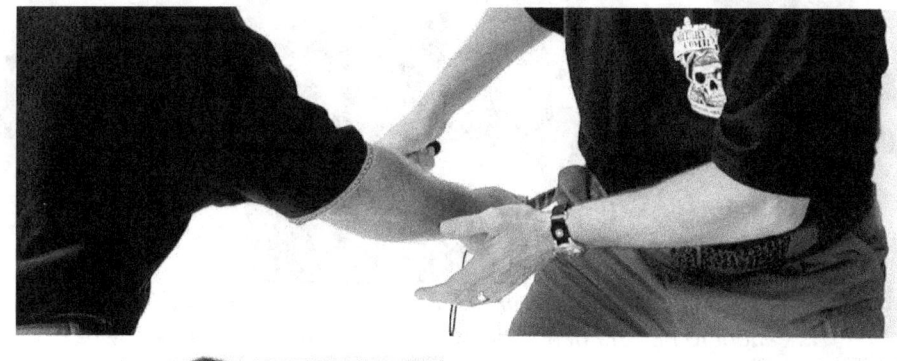

The soldier attempts a stab. The enemy slaps it aside. A mutual arm wrapping occurs.

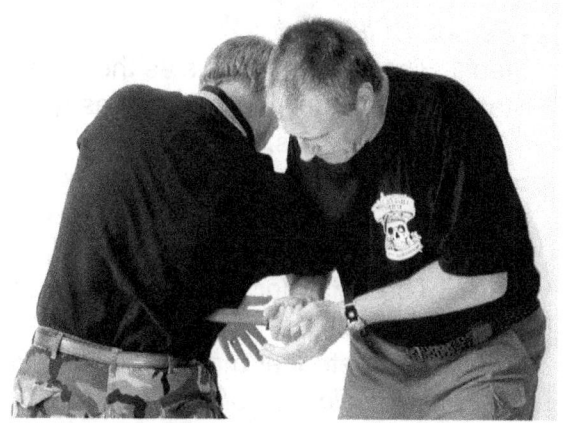

The soldier attempts a stab. The enemy slaps it aside. A mutual arm wrapping occurs.

Option Note: Should the enemy be bent forward, a kidney stab is an option to possibly draw the enemy upright. Otherwise - the soldier drives the point of his saber blade up into the base of the skull.

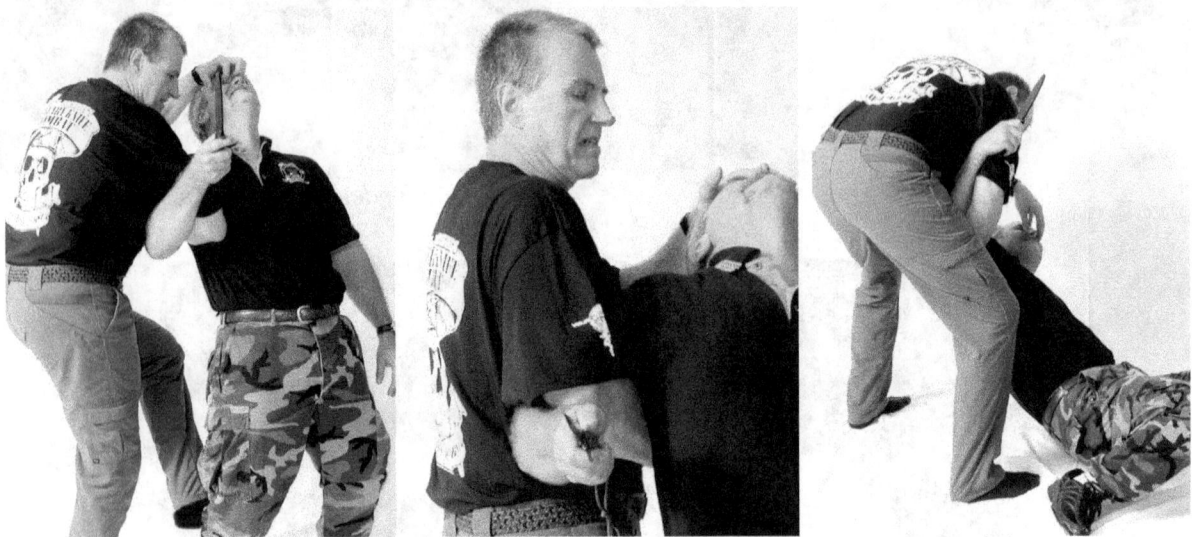

His fingers are buried into the enemy's eyes. The brow line is caught to pull the head backward and eventually, a take down.

2- Person Drill Practice:
  Step 1: The enemy attacks as the defender attacks) in low thrusting saber stab.
  Step 2: The enemy blocks, dodges and passes (or the defender blocks, dodges and passes.)
  Step 3: The defender and the enemy are in a mutual arm wrap trap.
  Step 4: The defender turns, reaches and grabs the forehead/brow of the enemy, with an eye attack.
  Step 5: The defender pulls the head back and drives the knife deep into the base of the skull.
  Step 6: The defender manipulates the knife to cause maximum damage.
  Note: *The soldier should take note of helmets, headgear or thick clothing worn by the enemy that might interfere with access to the base of the skull.*

Sample 4: Reverse Grip Stab to the Base of the Skull

Next in the progression is an isolated study in attacking the brain function through the base of the skull. The knifer attacks a vital nerve center in the base of the skull hoping to disrupt the cerebellum, and/or the medulla oblongata and/or perhaps even the top of the spinal cord. The knife should be pumped or twisted upon entry.

The enemy strikes. The defender blocks, pins and/or grabs the attacking limb.

The defender further pins by catching and cutting the enemy wrist, bloodline, and finger-control muscles.

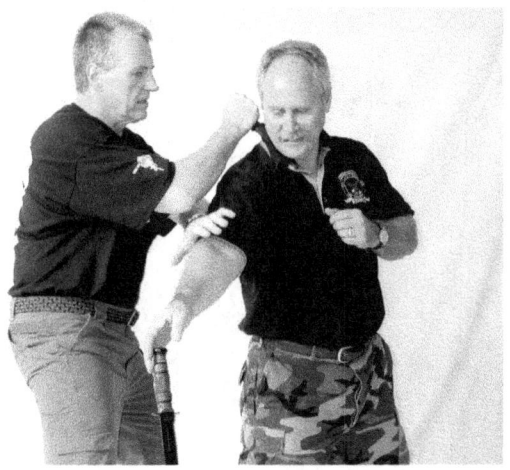

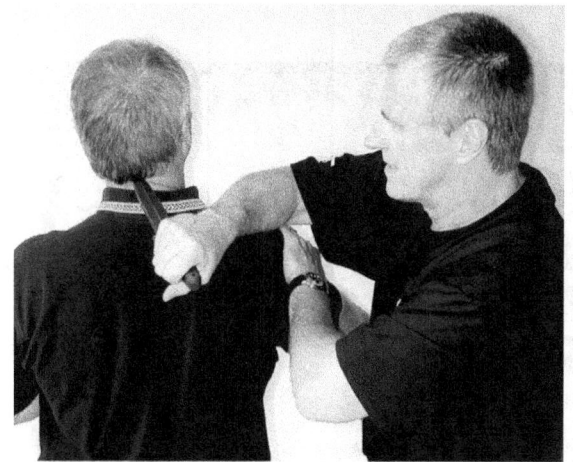

The defender slashes the enemy's carotid, pulling the knife far enough back to stab the base of the skull. Select any take down that suits your position.

2 Person Drill Practice:
- Step 1: The enemy attacks with any weapon, on the Combat Clock Basic 4 or the Advanced 12.
- Step 2: The defender blocks and pins the attack, cutting the wrist.
- Step 3: The defender slashes the enemy's carotid, then slips the knife back far enough to stab the base of the skull.
- **Note:** *The soldier should take note of helmets, headgear or thick clothing worn by the enemy that might interfere with access to the base of the skull.*

Sample 5: Saber Stab Uppercut through the Jaw Bone

Next in the learning progression is an isolated study of an uppercut attack, driven up through the chin, the mouth, roof of the mouth and nasal cavity into the brain. The knife must be long enough to reach this far. The knife must be long enough and slim, commando style.

The enemy attacks.

The uppercut. Once inside the jawbone, the tip must be driven in the right direction to achieve the brain target.

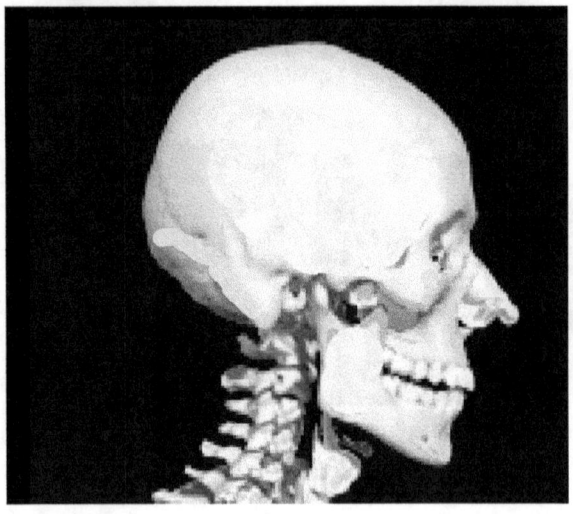

2 Person Drill Practice:

Step 1: The enemy attacks with any weapon, on the Combat Clock Basic 4 or Advanced 12.

Step 2: The defender stops the attack and stabs with an uppercut to the area under the chin.

Note: *This stab alone is horrible, brain target or not.*

Sample 6: Saber Stab the Throat and then the Notch

Next in the learning progression is an isolated study of an attack to the throat with a saber grip, this time with a horizontal blade to maximize damage. Consider the immediate reaction of the enemy faced with the gag factor and other reflexive and instinctive responses to such a stab. Should the defender's knife become stuck in the spinal cord, the defender should pump the handle sideways to maximize injury.

The defender raises his elbow, driving the horizontal blade down the throat maximizing the damage.

This knife is driven down into the clavicle notch and continued thrust turns into a takedown.

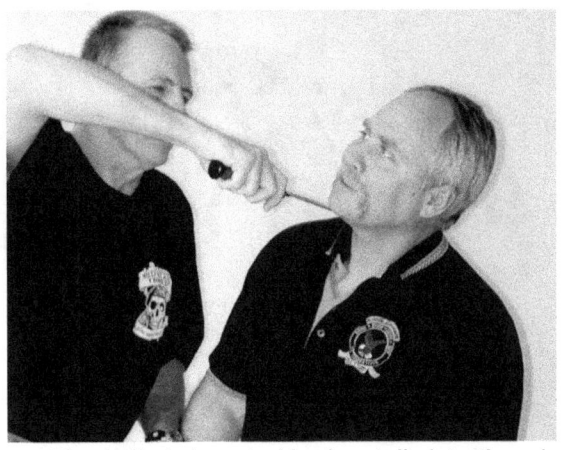
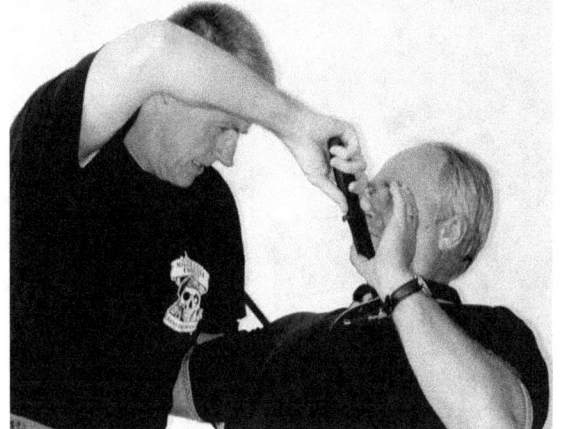

The knife is inserted horizontally into the windpipe. Raise the knife handle pushing the knife into the clavicle notch. Drive this down and forward into a take down.

2 Person Drill Practice:
 Step 1: The enemy attacks with any weapon, on the Combat Clock Basic 4 or Advanced 12.
 Step 2: The defender pins the attacking limb and power stabs the center of the throat with a horizontal blade.
 Step 3: The defender raises his elbow to drive the blade down into the throat. This continued downward pressure forces a takedown. The defender may or may not buckle the enemy's leg with a kick.
 Notes: If the knife is long, it may become stuck or restricted by the enemy's spinal cord. In missions requiring some level of silence, this stab may inhibit some vocal abilities of the enemy.

Sample 7: Reverse Grip Stab the Throat and then the Notch

Next in the learning progression is an isolated study of an attack to the throat with a reverse grip, this time with a horizontal blade to maximize damage. Consider the immediate reaction of the enemy faced with the gag factor and other reflexive and instinctive responses to such a stab. Should the defender's knife become stuck in the spinal cord, the defender should pump the handle sideways to maximize injury.

 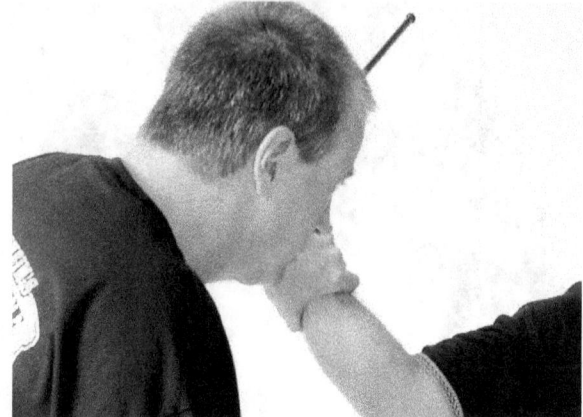

The enemy attacks and the defender defends, Starting with a block and grab.

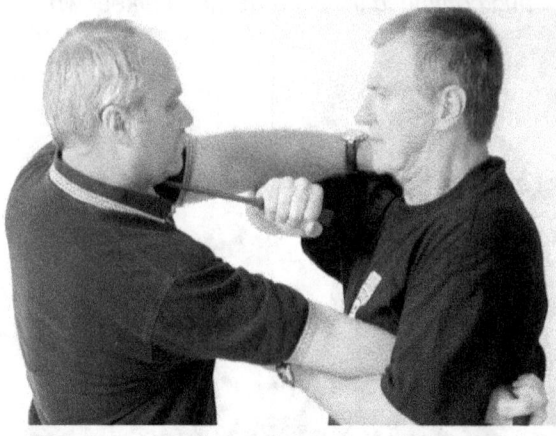 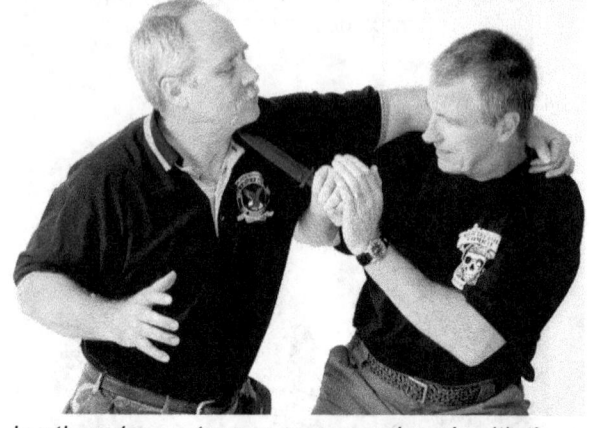

The defender stops, then wraps the weapon arm. The knife is driven into the throat.

In other circumstances, a pommel push with the support hand may help.

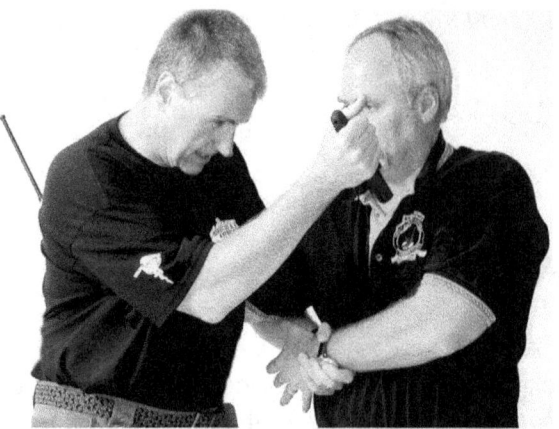 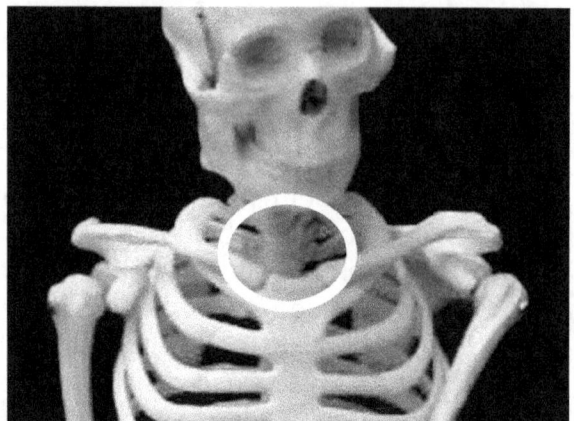

A possible take down is to drive the clavicle notch stab down and forward.

2 Person Drill Practice:
  Step 1: The enemy attacks with any weapon, on the Combat Clock Basic 4 or Advanced 12.
  Step 2: The defender draws a knife and charges.
  Step 3: The defender's arm wraps the weapon-bearing limb, grabs the enemy's free hand, and or then wraps the attacking limb, and stabs the enemy in his lower throat and clavicle area.
  Step 4: The defender pumps the handle sideways and/or twists the knife at the wrist and elbow.

Sample 8 Saber Slash to the Throat – "The Assassin's X"

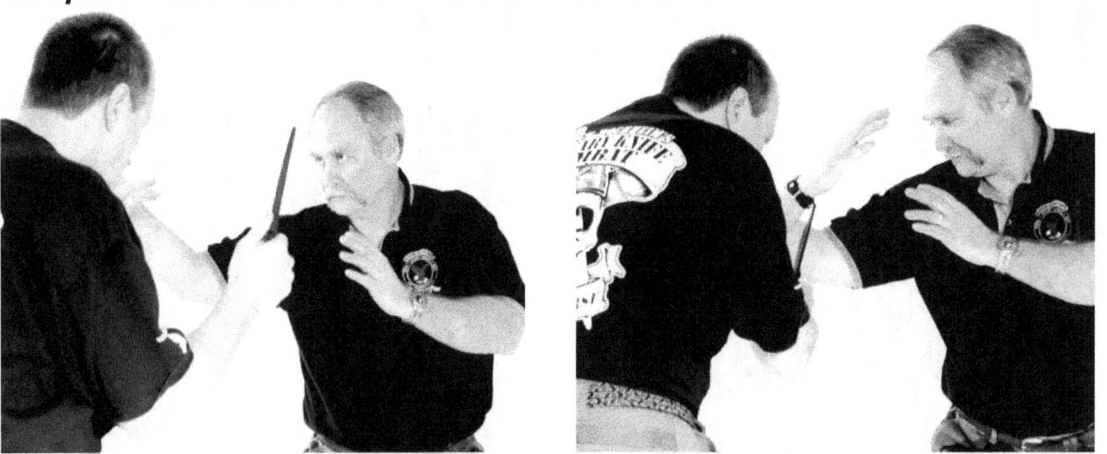

The enemy attacks with a baton. The soldier slashes/impacts the weapon bearing limb and hand attack the eyes.

Double slash the neck. The X-pattern begins. The defender slashes the neck (note the high left efficient attack). Then any aforementioned take downs.

2 Person Drill Practice
  Step 1: The enemy attacks with any weapon, on the Combat Clock Basic 4 or Advanced 12.
  Step 2: The defender draws a knife and charges, hitting the weapon-bearing limb and causing a disarm.
  Step 3: The defender X cuts the sides of the throat.

Sample 9: Saber Stab the Heart

Next in the learning progression is an isolated study of an attack to the heart with a saber grip stab. The human heart is a hollow, muscular organ about the size of a fist. It tends to be located more in the center of the chest than commonly thought, yet still somewhat over to the left side. The human heart is accessed from above the rib cage, through the rib cage, and from below the rib cage. Through the rib cage is best facilitated by somewhat of a horizontal blade. A power thrust may break ribs from any angled blade. Enter the chest but also aim somewhat to the center of the chest. Pump the knife handle sideways to maximize damage. The stab may also penetrate the lungs.

In this exercise, the enemy holds a pistol on the soldier within close range. The enemy becomes distracted and the defender draws a knife and charges. The defender's arm wraps the pistol bearing limb, grabs the enemy's free hand and stabs the enemy in his heart. The defender twists the knife at the wrist and elbow.

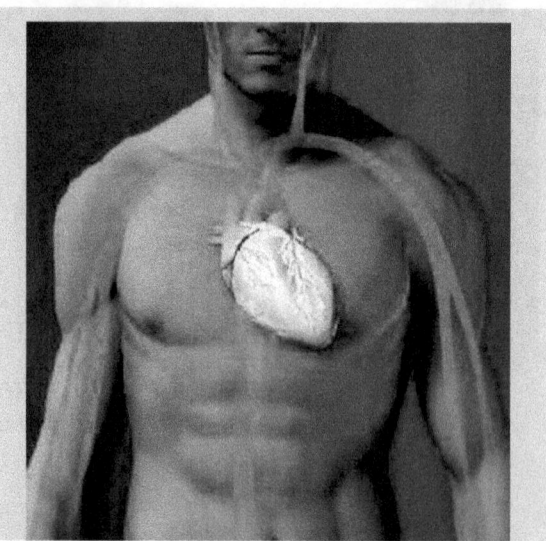

The human heart is a pear-shaped structure about the size of a fist. It is responsible for supplying the body with oxygenated blood. The heart lies in the chest cavity between the lungs.

2 Person Drill Practice:

Step 1: The enemy draws a pistol versus a soldier.
Step 2: The defender wraps the arm, draws a knife and charges.
Step 3: The defender grabs the enemy's free hand and stabs the enemy in his heart. The angle of the stab must facilitate entry into the rib cage.
Step 4: The defender twists the knife at the wrist and elbow, or pumps the handle sideways.
Note: Do not let the enemy free-fall down and back without controlling the pistol bearing limb. The defender must maintain the arm wrap; drop a knee if needed, to keep control of the pistol limb, else the enemy may still shoot the soldier in the throes of death.

Sample 10: Reverse Grip Stab the Heart

Next in the learning progression is an isolated study of an attack to the heart with an ice pick grip stab. The human heart is accessed from above the rib cage, through the rib cage, as best facilitated by a horizontal blade, and from below the rib cage. A power thrust may break ribs, no matter the angle of the knife. Missing the heart may mean a deadly stab and puncture into the lungs, causing damage to the enemy.

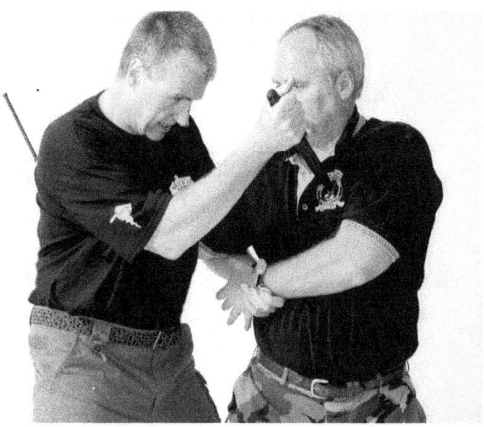

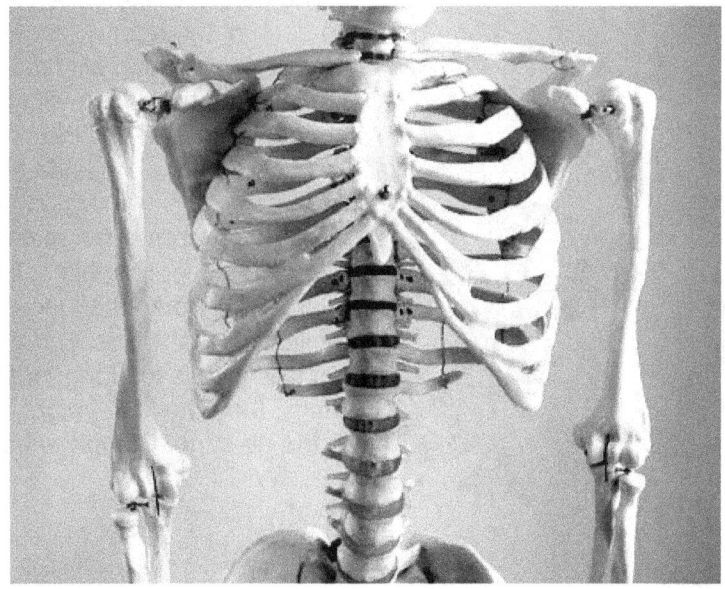

2 Person Drill Practice:

Step 1: The enemy attacks with any weapon, on the Combat Clock Basic 4 or Advanced 12.

Step 2: The defender draws a knife and charges.

Step 3: The soldier arm wraps the weapon-bearing limb, grabs the enemy's free hand and stabs the enemy in his heart. To facilitate the stab, the knife should be as horizontal as possible to enter through the ribs. A powerful, hammering stab will usually break the ribs.

Step 4: The defender pumps the handle sideways and/or twists the knife at the wrist and elbow.

Sample 11: The Saber Stab to the Diaphragm

Next in the learning progression is an isolated study of an attack to the solar plexus with a saber grip thrusting stab. The solar plexus is a dense cluster of nerve cells and supporting tissue, located just below the diaphragm. A blow or stab to that area, if it penetrates to the true solar plexus, not only causes great pain but may also temporarily halt visceral functioning like heartbeat and blood pressure. Follow up with any takedown.

 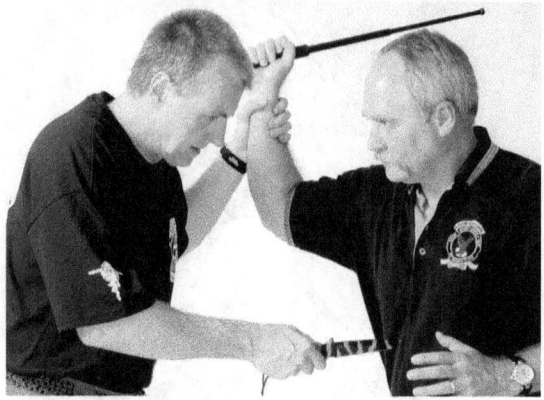

The defender stops the incoming attack and saber thrusts into the solar plexus.

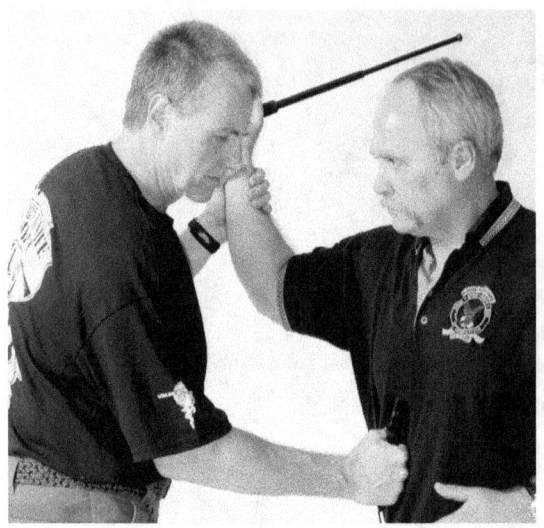

The defender then "1-inch punches" the stomach of the enemy, accentuating the bottom two knuckles of his fist. This powerful punching motion along with an assisted "uppercut punch" drive of the arm and shoulder. This should drive the knife blade up into the diaphragm.

The diaphragm is the major muscle of respiration. Piercing this balloon-like muscle that works the lungs is quickly debilitating to the enemy.

If the knife is too long, or the enemy too skinny, then the knife might become stuck on the rib cage. Pump the handle sideways.

2 Person Drill Practice

Step1: The enemy attacks with any weapon, on the Combat Clock Basic 4 or Advanced 12.

Step 2: The defender defends, blocks or catches the limb. He or she stabs the solar plexus.

Step 3: The defender aims the tip up. Using the "1 inch punch" concept, punches the two lower knuckle into the belly, helping the knife travel up into the diaphragm.

Note: *Though this is soft tissue it is neither hard nor easy either.*

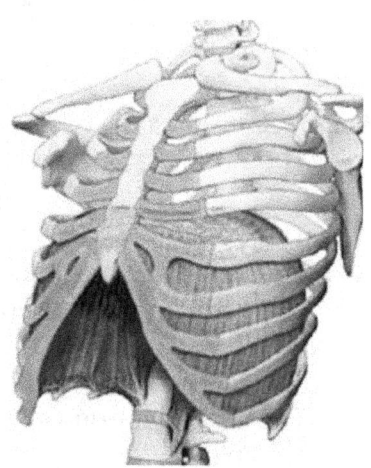

Sample 12: Saber Stab the Pelvis Region

Next in the learning progression is an isolated study of a saber grip, thrusting stab to the enemy's pelvis/groin. This is a physical and psychological injury. This is a medical center of nerves, muscles and veins. The enemy may become distracted and in an effort to avoid the incoming stab by bending, may be susceptible to a takedown with a continuous driving stab.

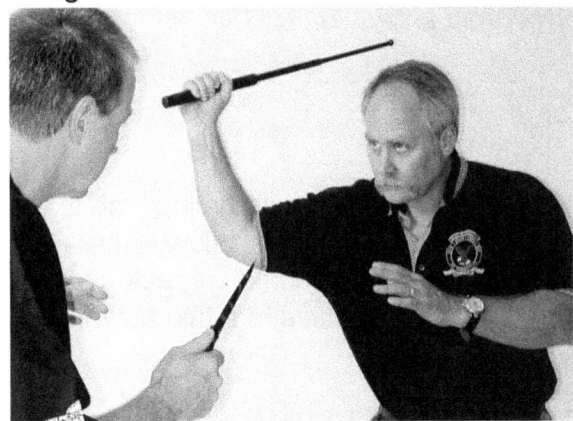

This enemy attacks.

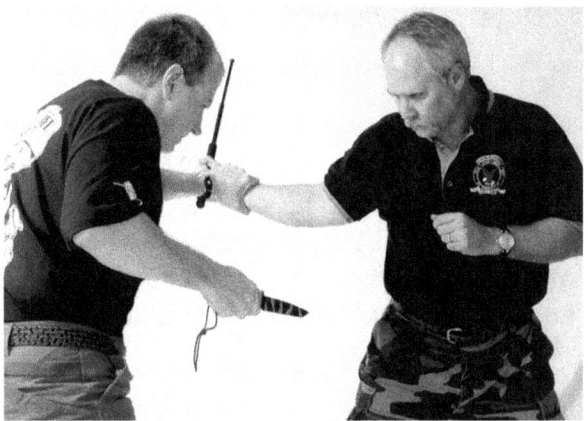

The defender stops the attack. The enemy is ready to block a high knife attack.

The defender fakes high and then stabs the pelvis.

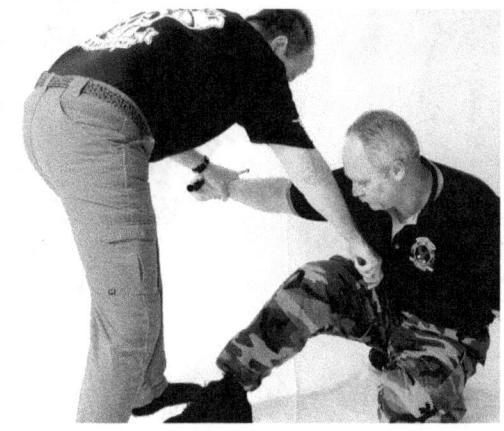

The defender drives the power stab downward and may actually force the enemy off balance.

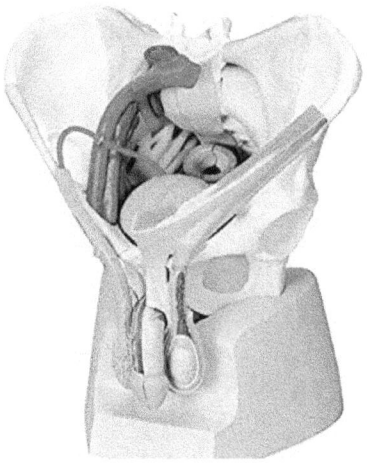

The pelvis is a very busy place. A medical center of muscle, blood, nerves and bone.

2 Person Drill Practice:
   Step 1: The enemy attacks.
   Step 2: The defender stops the attack and stabs the pelvis/groin in a powerful thrusting stab.
   Step 3: The defender may actually disrupt the enemy's balance with a downward power drive of the knife.
   Note: *The defender may hook stab the pelvis groin area.*

Sample 13: Saber Stab the Armpit
Next in the learning progression is an isolated study of a saber grip, thrusting stab to the enemy's armpit. As with the pelvis, this is a physical and psychological injury. This is a medical center of nerves, muscles and veins. The enemy may become distracted and in an effort to avoid the incoming stab by bending, may be susceptible to a takedown.

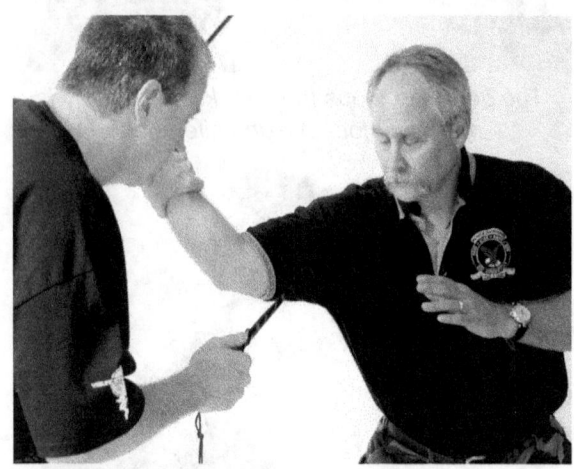
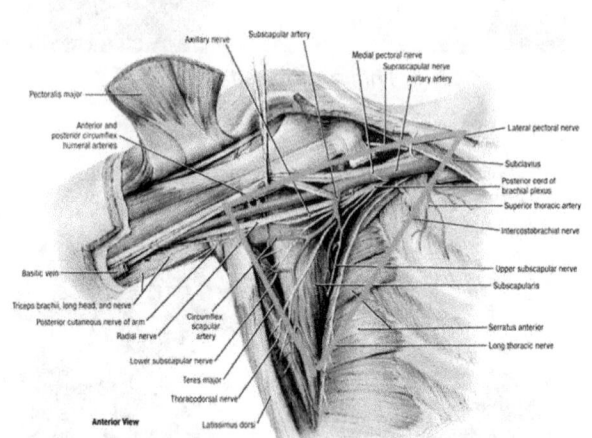

Suffice to say that there are a lot of things in the armpit.

2 Person Drill Practice:
   Step 1: The enemy attacks.
   Step 2: The defender stops the attack, stabs the armpit in a powerful thrusting stab.
   Step 3: The defender may twist and wrangle the injury. This may cause a weapon disarm.

Sample 14: Stabs to the Rear of the Heart – Military Classic #1

Next in the learning progression is an isolated study of an attack to the heart with a saber grip stab via the back. The human heart is best accessed by a horizontal blade. A power thrust with any angled blade may break ribs upon entry. A defender must be concerned about the shoulder blades and rib angles.

This rear, heart attack is a little known, little taught attack in most military circles. But, this power stab in the upper back may also pierce the lungs and damage the spinal cord, two other vital targets.

This was a scenario I learned in the Army, to face off with a classic military, trained knife-back stance enemy. Shown here with saber and reverse grips.

The defender sees that the enemy is distractible and subject to be tricked by feints and fakes.

The soldier fakes to his left and the enemy pivots to the perceived movement.

The defender stabs in a power, hooking manner such as this, planting the knife deep. If possible, pump the handle sideways to cause more injury.

The defender steps to the right and power plummets the knife into the back where he knows he might reach the heart. The angle of the knife fits the ribs.

The defender stabs in a power, hooking manner such as this, planting the knife deep. If possible, pump the handle sideways to cause more injury.

Targeting the common location of the heart is important, along with the size of your knife. But, it would be possible that your knife might become lodged in the enemy's rib cage and you might not have the chance to properly extract the knife. When this happens, in old military lore – "if you can't get your knife out, he can't get it out either." In which case, a suggested option is to let loose of the handle of your embedded, stuck knife and fight on with the enemy, using the empty hand combatives of severe striking, kicking and grappling. Once the enemy is incapacitated, "then get your knife."

Knife stuck in bone? Let it go and one stunning option is a palm-heel strike to the base of the skull. This power strike should splash the brain inside the skull and is known to be effective. The shock may even cause the enemy to drop the knife, if the stab doesn't.

The reverse grip version. Next in the learning progression is an isolated study of a ice pick/reverse grip stabbing attack to the heart through the back. Here, the human heart is best accessed by a horizontal blade running in an angle parallel with the rib cage, but a power thrust with any angled blade may break these ribs upon entry. Read the information about the human heart in the prior segment. It is important to note that the skeletal structure of the back is different than the front and the heart has more bone protection from rear attack. Military manuals call them the shoulder blades. Doctors call them the scapula. They are high and off to the sides on the torso, and while not invincible from knife penetration, they still offer some level of shielding. To make this stab, a soldier needs to be aware of the scapula and the rib cage, and the common angle of ribs and common space between ribs. Concerns about the scapula are often enough to dissuade soldiers from attempting this stab.

 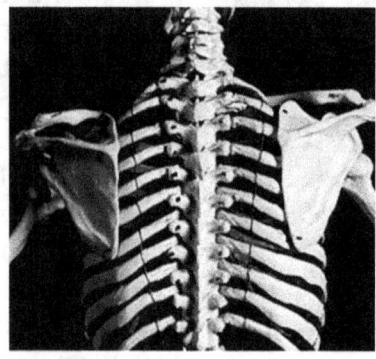 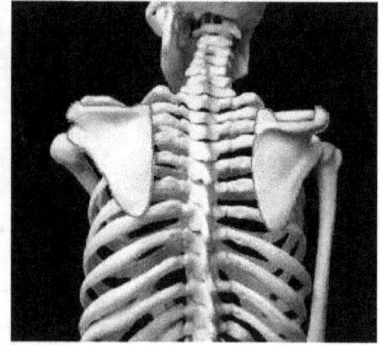

Scapulas vary somewhat in size and thickness person to person, and the soldier must be aware of the scapula and the common angles of the rear rib cage to access the heart. These bones also can be penetrated or broken with a targeted power stab. More on this anatomy in the subsequent segment. If the attack misses the heart, the soldier may still injure the lungs or spinal cord.

The enemy attacks and the soldier responds to see an open target. Knowledge of anatomy guides the stab between the scapula and the spinal cord with the knife tip aimed at the left center of the chest.

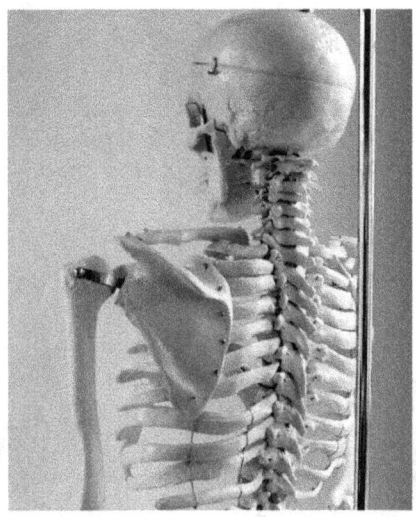 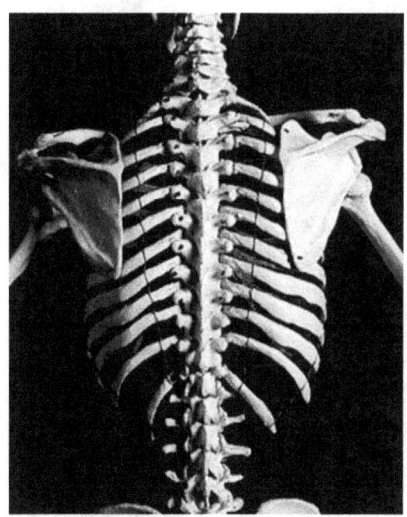

2 Person Drill Practice:
   Step 1: The soldier and the enemy stand-off. The enemy stands in the classic military "knife-back" stance.
   Step 2: The soldier fakes to his left, drawing the attention of the enemy, who pivots his torso.
   Step 3: The soldier power stabs the enemy's left back, where he might access the heart from the rear.
   Step 4: The soldier pulls and pushes the handle in a pumping fashion to maximize pain.
   Note: *It is possible that the knife may become wedged into the rib cage or scapula. If so, leave it and finish the fight with your empty hands.*

Sample 15: The Three Uppercuts – Military Classic #2

Next in the learning progression is an isolated study of uppercut attacks, taken in a practical progression. This is taught here in a series of possible targets, as the enemy might dodge the uppercut to the groin, or dodge an uppercut to the stomach or an uppercut to under the chin. This was shown to me by South African Commando Captain and war vet Ben Mangels.

The enemy attacks.

The soldier blocks and grabs and then uppercuts the groin.

The enemy bends back to save his groin, exposing his stomach.

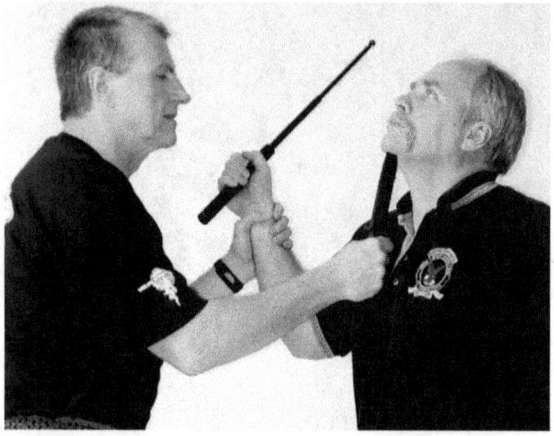

The enemy backs away from the stomach groin stab and exposes the throat under the chin.

2-Person Drill Practice:
   Step 1: The enemy attacks the soldier.
   Step 2: The soldier stops the attack, as access to the enemy's "center line," and stabs:
            Target 1: An uppercut to the groin, or,
            Target 2: An uppercut to the stomach or,
            Target 3: An uppercut to the area under the chin as the head might extend as a bending dodge.
   Step 3: Execute any takedown and appropriate finish.

Sample 16: Escape a Rear Choke Ambush
If the choke is amateurishly a bit sloppy and you have some time. Here is a knife solution.

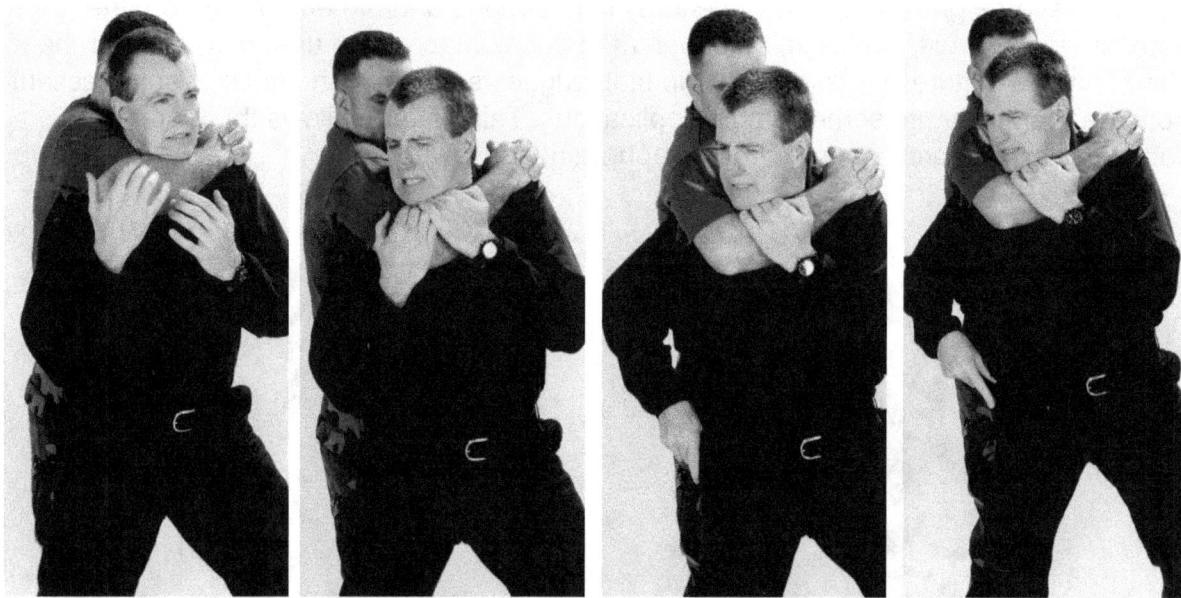

The soldier is jumped. Seize the arm and maneuver your neck until the airway is at least partially open. There are several solutions, but a knife solution is chosen. Access the folding knife of a fixed blade knife.

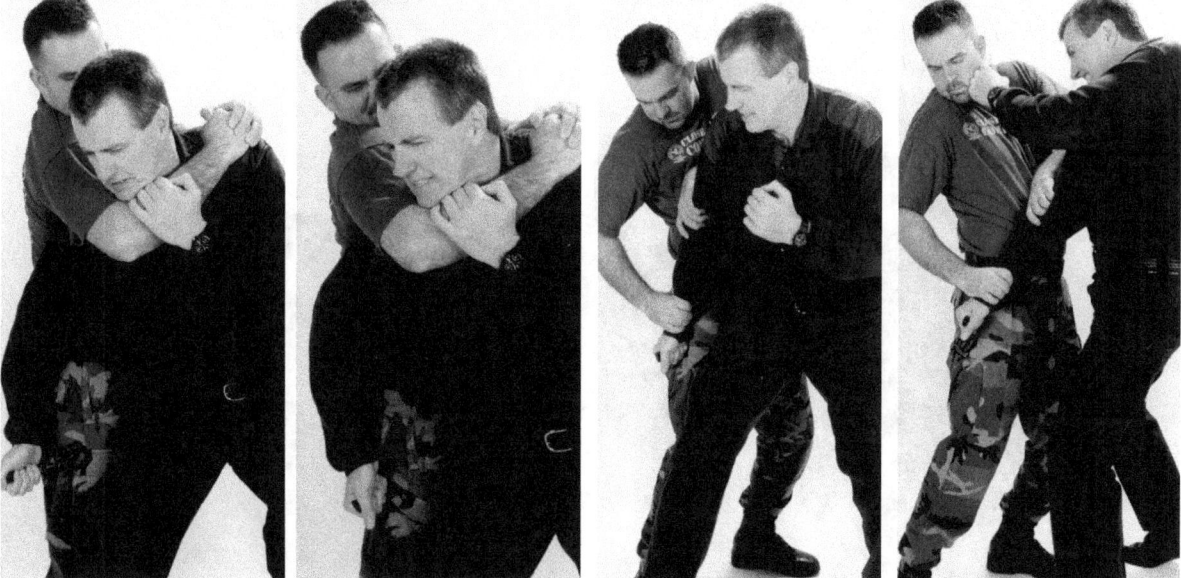

Stab the side of the choking arm multiple times until the enemy loosens the choke in an effort to stop the destruction of his leg. The soldier beats the face and neck of the enemy. He slap releases the enemy grip and rabidly uses his knife.

NEVER, ever stab the arm around your neck! It's right by your neck! You will be in moving, chaotic, struggling motion and could kill yourself!

Next in the rear choke escape, learning progression is an isolated study of ice pick, stabbing uppercut attacks, taken in a practical progression. This is taught here in a series of possible reverse grip targets, as the enemy might dodge and/or bend away from the uppercut to the groin, or dodge an uppercut to the stomach or an uppercut to under the chin. The very nature of this reverse grip stab suggests an ambush attack or some assault from the rear. Hear are some sample applications. This pattern follows the aforementioned military uppercut series. Groin, diaphragm, and jawbone.

This angle of reverse grip attack is best suited for an attack from the rear. Here, the leg is stabbed.

A groin stab.

The enemy bends back to save his groin, exposing his stomach. Or, the soldier just targets the stomach.

The enemy backs away from the groin or stomach stab and exposes his throat under the chin.

2 Person Practice Drill
  Step 1: The attacker grabs the defender from behind.
  Step 2: The defender deals with the choke, if choked, then responds with a knife.

Sample 17: Stand-off Duel Solution – Military Classic #3

This was shown to me by South African Commando Captain and War Vet Ben Mangels.

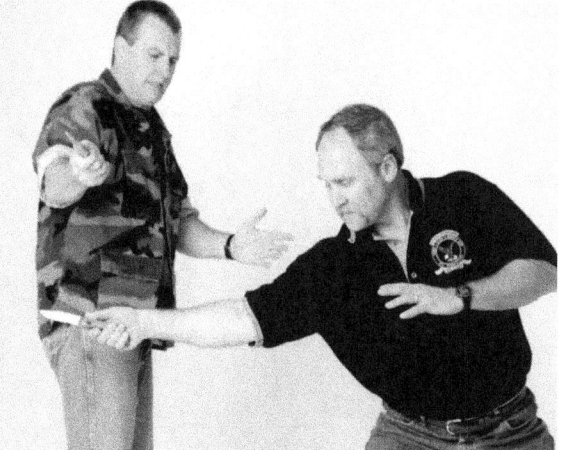

The knife face-off duel. The enemy power lunges and the soldier dodges.
His knife arm is above the enemy's knife arm. This may well cause an impact disarm.

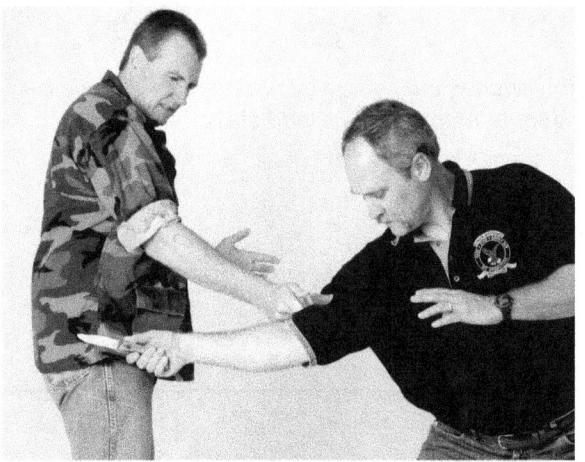

The soldier strikes the arm as deeply as possible.   *The soldier next strikes the enemy's neck.*

The soldier is now behind the enemy. Monitoring the enemy's
torso, he shoves the knife into the enemy's kidney area.

 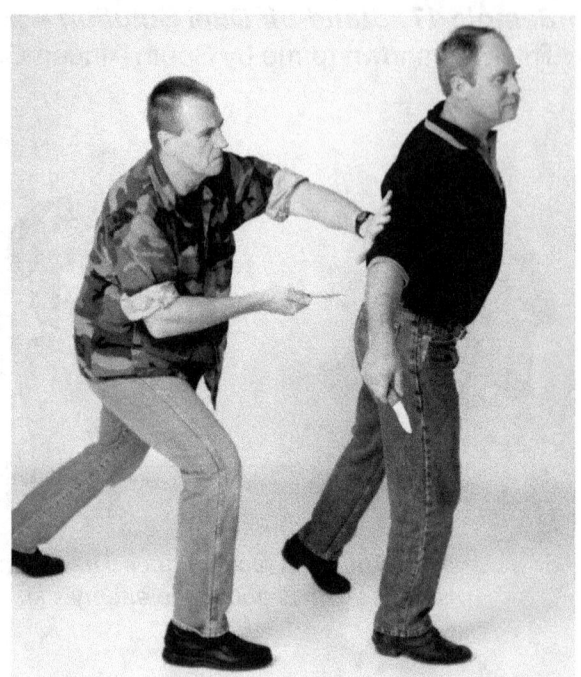

The soldier positions his arm (or shoulder) against the enemy, pushes away and yanks the knife out. From this position the soldier may escape, or execute more slashing and stabbing attacks.

Black Box Knife Combat Files

"The two stood off-Rai with his kukri and the Chinese with his knife. The Chinese started moving his hands in circles like a boxer, his front hand empty and open, his knife back near his chest. Rai leapt off and hit right on the front arm with that big curved blade. It [the arm] almost fell off near the elbow! But before blood could even spew, in an instant, Rai swung hard at the man's head. The end of his kukri tore into the Chinese's neck, and Chinaman fell into a human pile on the ground. I swear to all that is Holy, it was that fast. Two seconds. Two swings. Arm gone. Neck gone. A man gone.

His [Rai's] associates dragged the man off the platform, not once removing their eyes from Rai. A slipper remained on the floor, raked from a foot – a slipper and the dead man's knife. Rai was like that, sure as bloody hell a man to contend with."

– Len Archibald,
The London Times, England

Sample 18: Using The Releases – Cut 'em and twist 'em
These are best expressed through predicaments from chaotic combat scenarios. Sample of clutches combat scenarios follow. All ground fighting strategies will be documented in the subsequent ground, *Rattlesnake* segments.

  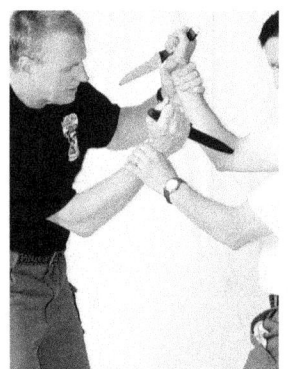 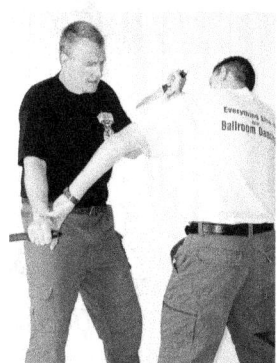

The high clutches.   Defender attacks grabbing limb and attacks the knife limb.   Defender uses a circular release.

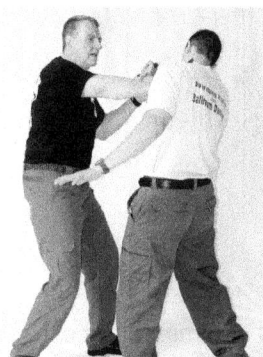  

Throat slash to...   solar plexus stab to groin stab, causing a body bend...   to bent body hooked...

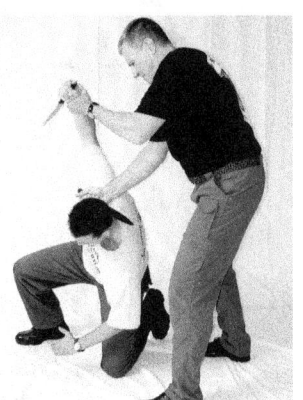 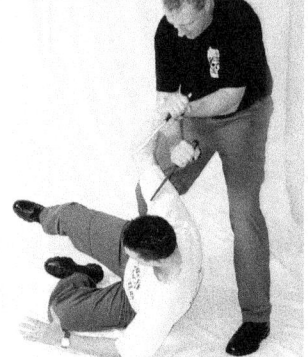

...into a torso wheel throw.

Sample 19: The Pommel Charge

The high clutch.

Defender attacks and weakens the enemy's grip.

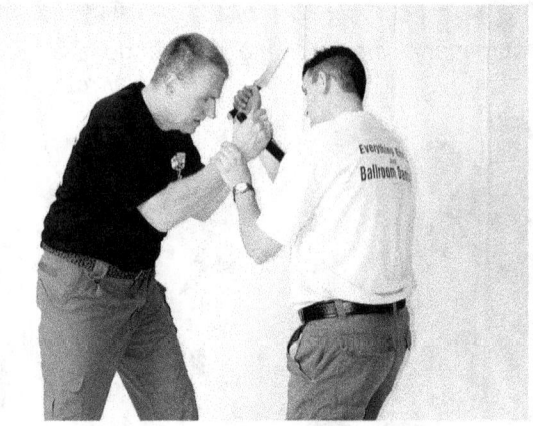

The defender stabs the enemy's knife arm.

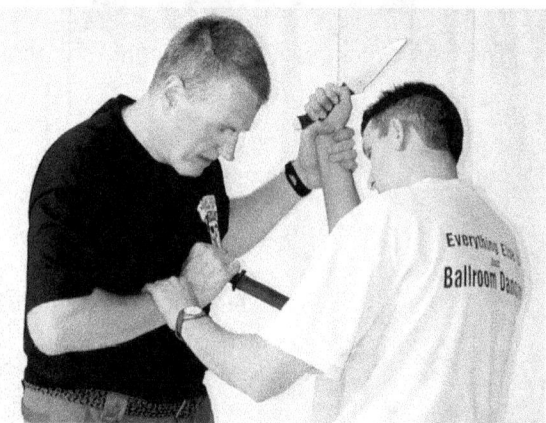

The defender braces his pommel on his chest (heart or solar plexus).

The defender charges forward. Once stab is in, the defender may cant the knife upward into the lungs. A turn downward may cause the enemy to collapse down.

Any number of trips and leg sweeps could be used. The defender should try to yank his knife back out of the enemy.

Sample 20: Wrap and Defeat the Arm Grab

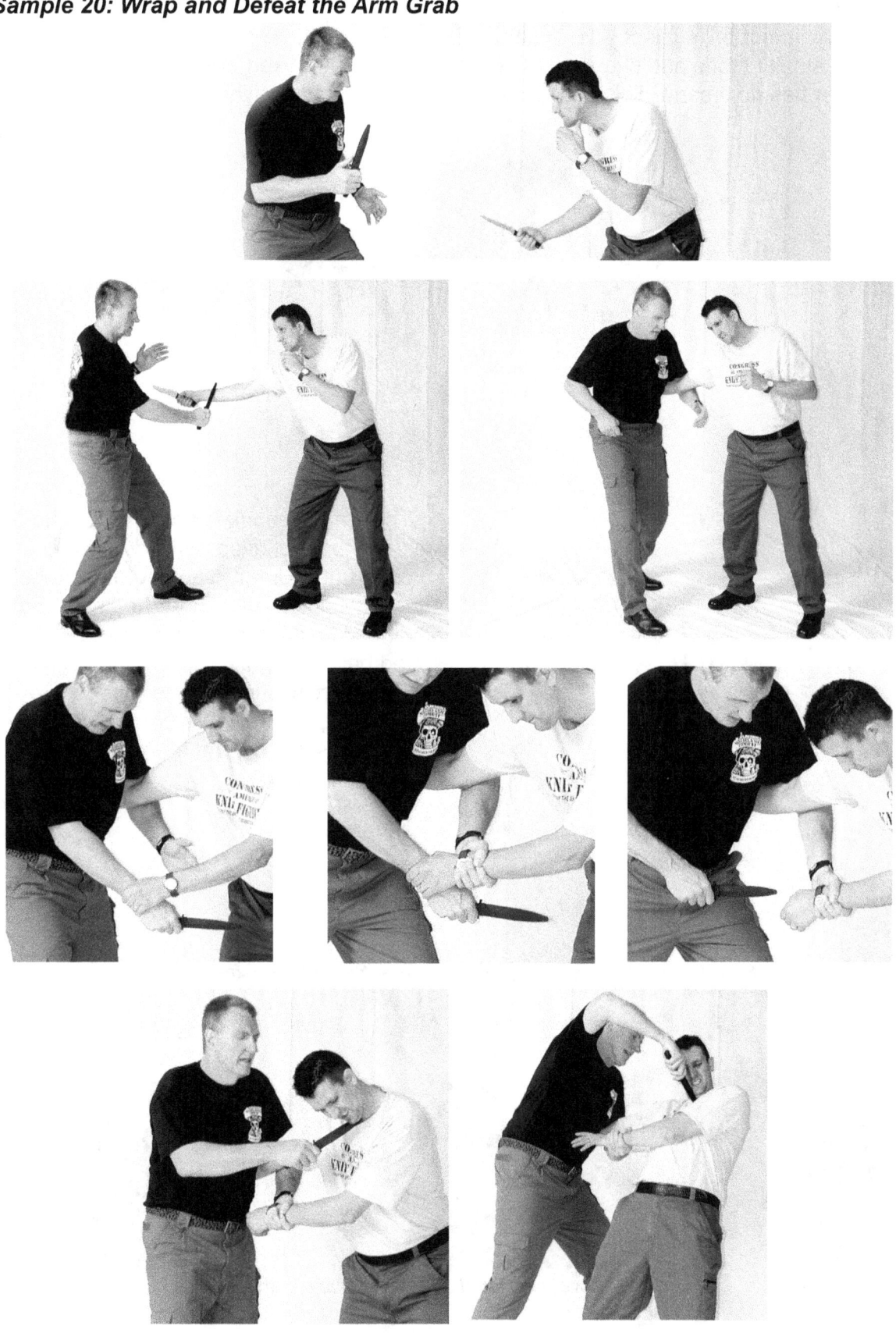

Sample 21 – Study Subject: Knife versus Bayonet

This is an important subject, little studied if at all. The enemy is out of ammo or must remain in stealth mode and should not fire. Or, they are intermixed with friendlies. The defender has no firearm. This is a military problem but then, maybe not...

This training book is about hand held knife combat not rifled, affixed bayonet combat, but in order to defeat the bayonet attack, a soldier must know the basic and advanced training of the unskilled and skilled bayonet fighter.

The exercises listed here will prepare the soldier to fight bayonet armed enemy. For comprehensive and further training, review the attack list below.

Basic Training Step 1: The 12 o'clock High Encounter – Saber Grip Example

Move, block and grab the gun and stab.

Basic Training Step 2: The 3 o'clock Right-side Encounter – Saber Grip Example

Move, block, grab the gun and stab.

Basic Training Step 3: The 6 o'clock Low Encounter – Saber Grip Example

Move, block, grab the gun and stab.

Basic Training Step 4: The 9 o'clock Left-side Encounter – Saber Grip Example

Move, block, grab the gun and stab.

This grab upon the weapon is key. The enemy must not be able to retract his stab, or raise his rifle to block your stab. Seize the weapon with dedicated strength.

The grab upon the enemy's incoming rifle is important. The enemy will probably maintain a two-handed grip on his long gun. The soldier must grab and push the weapon in such manner as to secure an opening for the knife stab to pierce into a key target. The following are examples of these encounters.

The Basic Training Bayonet Stab Set
The enemy stabs at 12 o'clock – the defender moves, blocks, grabs the gun and stabs.
The enemy stabs at 3 o'clock – the defender moves, blocks, grabs the gun and stabs.
The enemy stabs at 6 o'clock – the defender moves, blocks, grabs the gun and stabs.
The enemy stabs at 9 o'clock – the defender moves, blocks, grabs the gun and stabs.
The enemy stabs at the center – the defender moves, blocks, grabs the gun and stabs.
Note: *You defend against these attacks while knee-high and the enemy is standing.*
      You defend against these attacks while on your back and the enemy is standing.

The Advanced Training Bayonet Stab Set
The enemy stabs at 12 o'clock – the defender moves, blocks, grabs the gun and stabs.
The enemy stabs at 1 o'clock – the defender moves, blocks, grabs the gun and stabs.
The enemy stabs at 2 o'clock – the defender moves, blocks, grabs the gun and stabs.
The enemy stabs at 3 o'clock – the defender moves, blocks, grabs the gun and stabs.
The enemy stabs at 4 o'clock – the defender moves, blocks, grabs the gun and stabs.
The enemy stabs at 5 o'clock – the defender moves, blocks, grabs the gun and stabs.
The enemy stabs at 6 o'clock – the defender moves, blocks, grabs the gun and stabs.
The enemy stabs at 7 o'clock – the defender moves, blocks, grabs the gun and stabs.
The enemy stabs at 8 o'clock – the defender moves, blocks, grabs the gun and stabs.
The enemy stabs at 9 o'clock – the defender moves, blocks, grabs the gun and stabs.
The enemy stabs at 10 o'clock – the defender moves, blocks, grabs the gun and stabs.
The enemy stabs at 11 o'clock – the defender moves, blocks, grabs the gun and stabs.
Note: *You defend against these attacks while knee-high and the enemy is standing.*
You defend against these attacks while on your back and the enemy is standing.

The Basic Training Set Bayonet Slash Set
The enemy slashes at 12 o'clock – the defender moves, blocks, grabs the gun and stabs.
The enemy slashes at 3 o'clock – the defender moves, blocks, grabs the gun and stabs.
The enemy slashes at 6 o'clock – the defender moves, blocks, grabs the gun and stabs.
The enemy slashes at 9 o'clock – the defender moves, blocks, grabs the gun and stabs.
The enemy slashes at the center – the defender moves, blocks, grabs the gun and stabs.
Note: *You defend against these attacks while knee-high and the enemy is standing.*
You defend against these attacks while on your back and the enemy is standing.

The Advanced Training Set Bayonet Slash Set
The enemy slashes at 12 o'clock – the defender moves, blocks, grabs the gun and stabs.
The enemy slashes at 1 o'clock – the defender moves, blocks, grabs the gun and stabs.
The enemy slashes at 2 o'clock – the defender moves, blocks, grabs the gun and stabs.
The enemy slashes at 3 o'clock – the defender moves, blocks, grabs the gun and stabs.
The enemy slashes at 4 o'clock – the defender moves, blocks, grabs the gun and stabs.
The enemy slashes at 5 o'clock – the defender moves, blocks, grabs the gun and stabs.
The enemy slashes at 6 o'clock – the defender moves, blocks, grabs the gun and stabs.
The enemy slashes at 7 o'clock – the defender moves, blocks, grabs the gun and stabs.
The enemy slashes at 8 o'clock – the defender moves, blocks, grabs the gun and stabs.
The enemy slashes at 9 o'clock – the defender moves, blocks, grabs the gun and stabs.
The enemy slashes at 10 o'clock – the defender moves, blocks, grabs the gun and stabs.
The enemy slashes at 11 o'clock – the defender moves, blocks, grabs the gun and stabs.
Note: *You defend against these attacks while knee-high and the enemy is standing.*
You defend against these attacks while on your back and the enemy is standing.

The Bayonet Attack Exercise Drill

1) The enemy thrust stabs on the combat clock (standing and ground).
2) The enemy hooking slashes on the combat clock.
3) The enemy thrusting, long gun butt strikes on the combat clock.
4) The enemy hooking long gun butt strikes on the combat clock.
5) The enemy strikes in combination attacks:
   Example a) thrusting butt strike, hooking bayonet slash.
   Example b) hooking bayonet slash and thrusting butt strike.
   Example c) thrusting bayonet stab and hooking butt stroke.
   Example d) combinations.

6) The defender blocks, grabs the weapon and stabs or slashes the enemy in key target areas.
7) The defender completes a combat scenario with takedowns, wary of the bayonet and rifle barrel position.

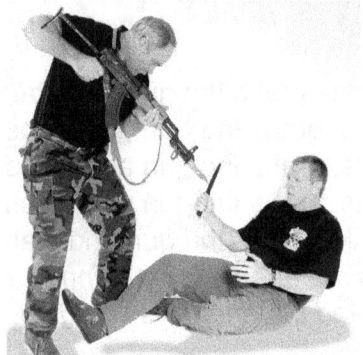 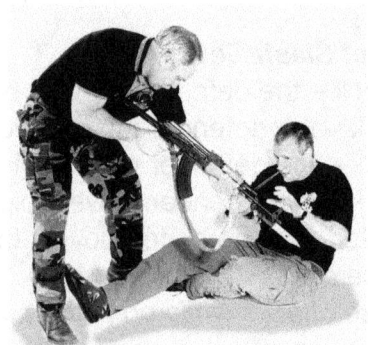 

Block, grab if possible. Try to get up!

Hugg underifrån

Grip motståndarens vapen med vänster hand och för det åt sidan

Håll kvar motståndarens vapen och hugg med full kraft

Hugg underifrån

Grip motståndarens vapen med vänster hand och för det åt sidan

Håll kvar motståndarens vapen och hugg med full kraft

Hugg uppifrån

Hugg uppifrån

Sample 22 – Study: Knife Sentry Killing

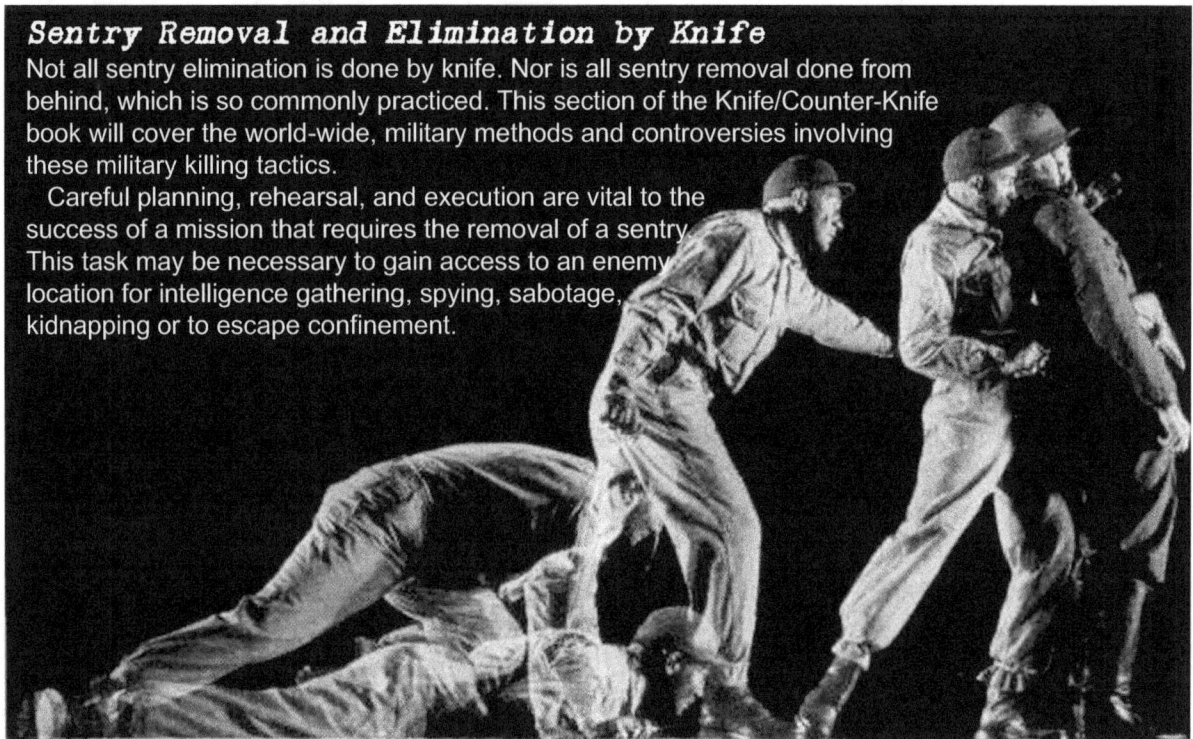

Sentry Removal and Elimination by Knife
Not all sentry elimination is done by knife. Nor is all sentry removal done from behind, which is so commonly practiced. This section of the Knife/Counter-Knife book will cover the world-wide, military methods and controversies involving these military killing tactics.

Careful planning, rehearsal, and execution are vital to the success of a mission that requires the removal of a sentry. This task may be necessary to gain access to an enemy location for intelligence gathering, spying, sabotage, kidnapping or to escape confinement.

Psychological Aspects

Killing a sentry is completely different than killing an enemy soldier while engaged in a firefight. It is a cold and calculated attack on a specific target. After observing a sentry for hours, watching him eat or look at his wife's photo, an attachment is made between the stalker and the sentry. Nonetheless, the stalker must accomplish his task efficiently and brutally. At such close quarters, the soldier literally feels the sentry fight for his life. The sights, sounds, and smells of this act are imprinted in the soldier's mind; it is an intensely personal experience. A soldier who has removed a sentry should be observed for signs of unusual behavior for four to seven days after the act.

Planning the Attack

Military experts suggest that a detailed schematic of the layout of the area guarded by sentries must be available. Mark known and suspected locations of all sentries. It will be necessary to learn the schedule for the changing of the guards and the checking of the posts. These methods include plans to:

^ learn the guard's meal times. It may be best to attack a sentry soon after he has eaten when his guard is lowered. Another good time to attack the sentry is when he is going to the latrine.
^ attack in the a.m. hours, such as 4 a.m. when sentry is biologically tired and sleepy.
^ develop a contingency plan.
^ to plan infiltration and exfiltration routes.
^ to carefully select personnel to accomplish the task.
^ to carry the least equipment necessary to accomplish the mission because silence, stealth, and ease of movement are essential.

^ to conceal or dispose of killed sentries.

Rehearse the Attack
Reproduce and rehearse the scenario of the mission as closely as possible to the execution phase. Conduct the rehearsal on similar terrain, using sentries, the time schedule, and the contingency plan. Use all possible infiltration and exfiltration routes to determine which may be the best.

Execute the Attack
When removing a sentry, the soldier uses his stalking skills to approach the enemy undetected. He must use all available concealment and keep his silhouette as low as possible.

Execution 1: When stepping, the soldier places the ball of his lead foot down first and checks for stability and silence of the surface to be crossed. He then lightly touches the heel of his lead foot. Next, he transfers his body weight to his lead foot by shifting his body forward in a relaxed manner. With the weight on the lead foot, he can bring his rear foot forward in a similar manner.

Execution 2: When approaching the sentry, the soldier synchronizes his steps and movement with the enemy's, masking any sounds. He also uses back ground noises to mask his sounds. He can even follow the sentry through locked doors this way. He is always ready to strike immediately if he is discovered. He focuses his attention on the sentry's head since that is where the sentry generates all of his movement and attention.
    However, it is important not to stare at the enemy because he may sense the stalker's presence through a sixth sense? He focuses on the sentry's movements with his peripheral vision. He gets to within 3 or 4 feet and at the proper moment makes the kill as quickly and silently as possible.

Execution 3: The attacker's primary focus is to summon all of his mental and physical power to suddenly explode onto the target. He maintains an attitude of complete confidence throughout the execution. He must control fear and hesitation because one instant of hesitation could cause his defeat and compromise the entire mission.

Knife Tactics
The following techniques are proven and effective ways to remove sentries. A soldier with moderate training can execute the proper technique for his situation, when he needs to. The following are the common, military tactics. Soldiers across the globe argue about which ones are right or wrong, better or worse:

Will the soldier also stab his own arm here?

Mouth cupped for silence and chin lifted.

Saber grip is pulled across the throat in a slashing manner, or...

Saber grip knife stabbed into the side of throat and pushed forward, or...

Saber grip is stabbed into the base of the skull, or...

Saber grip is stabbed into a kidney, or...

Reverse grip into throat and clavicle, or...

**Reverse grip slash across the throat, or...
Pommel strike for a stunning blow, then follow with stabs and/or slashes.**

Controversies?

Soldiers from various countries and some from within the same country will argue that some of the sentry kill stabs and slashes are dangerous and some will even declare them "wrong!" Most of the arguments stem from the forgotten reality that when you snatch a sentry with your support arm, often a brief tussle may ensue and the sentry may well be moving violently about. His energized movement causes the soldier to also move about. If a soldier slashes the neck? The soldiers are essentially pulling the knife edge back at themselves. This accident is a possibility.

As an alternative naysayers suggest stabbing the side of the throat and "punching" the knife forward, with the intention of ripping the windpipe out and anything else in the path with it. Other naysayers of this method complain that in the "struggling sentry model" the sentry is struggling and should a stabber miss the stab and punch the knife forward in one trained "stab and punch" move? The soldier will reflexively raise a hand and grab the weapon-bearing forearm. This is also a possibility.

A Vietnam veteran reported to me once that he once stabbed a enemy guard in the side of the neck. The knife went through the thin neck and he stabbed his own left support forearm that held the VC's face and mouth. This soldier instinctively yanked that arm back and cut his own forearm open for several inches with the tip of his knife, protruding from the VC's neck.

The same is said of the various reverse grip slashes and stabs. The reverse grip slash may come back upon the soldier, cutting him. The somewhat vertical, downward stabs may also hit the soldier as the sentry moves energetically about after a sudden surprise. These are also possibilities.

 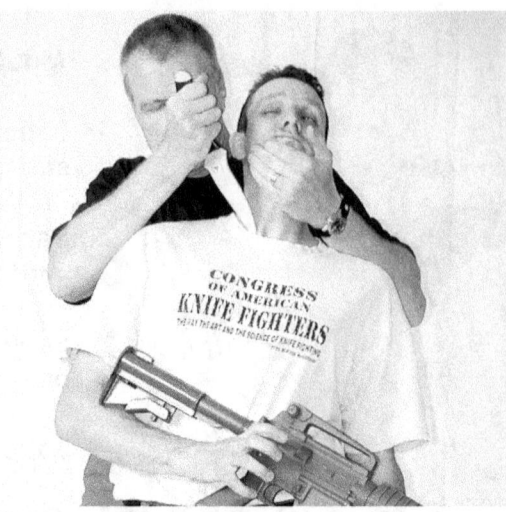

Naysayers will argue that stabs or slashes, saber or ice pick grip can be so dangerous that some classic moves are wrong moves. Yet, all the attacks have faults against an energized, moving sentry.

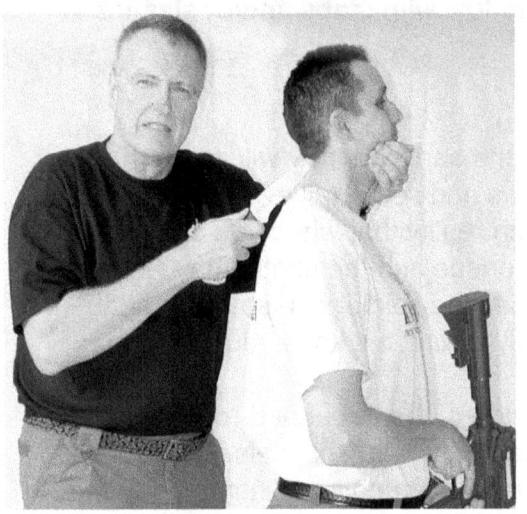

Even the twisting blade to the base of the skull, against a struggling sentry might cause a cut to the soldier's support forearm.

Many older military manuals teach the kidney stab as a quick, sentry kill. Some of these manuals claim that the stab is so painful and deadly that the sentry cannot even utter a sound and dies quickly. Yet, veterans and forensic experts declare the kidney stab is not so quiet and not always so successful.

Classical Military Sentry Knife Kill Scenarios
The following few combat scenarios come straight from military manuals.

Stun and Throat Cut
This technique relies on complete mental stunning to enable the soldier to cut the sentry's throat, severing the trachea and carotid arteries. Death results within 5 to 20 seconds. Some sounds are emitted from the exposed trachea, but the throat can be cut before the sentry can recover from the effect of the stunning strike and cry out. The soldier silently approaches to within striking range of the sentry as shown in photos here.

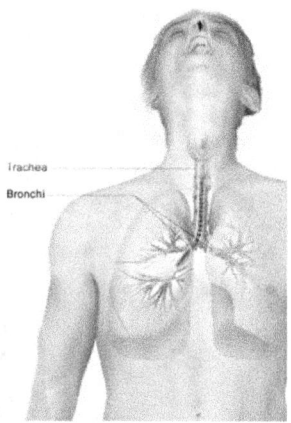

Step 1. The soldier strikes the side of the sentry's neck with the knife pommel/ butt as shown in Step 2, which completely stuns the sentry for three to seven seconds. He then uses his body weight to direct the sentry's body to sink in one direction and uses his other hand to twist the sentry's head to the side, deeply cutting the throat across the front in the opposite direction as in Step 3. He executes the entire length of the blade in a slicing motion. The sentry's sinking body provides most of the force, not the soldier's upper-arm strength as in Step 4.

Kidney Stab, Throat Cut.

This is a very popular method. The kidneys are located behind the abdominal cavity. There are two, one on each side of the spine and they are approximately at the vertebral level T12 to L3. The right sits just below the diaphragm and posterior to the liver, the left below the diaphragm and posterior to the spleen. Above each kidney is an adrenal gland. Parts of the kidneys are partially protected by the eleventh and twelfth ribs.

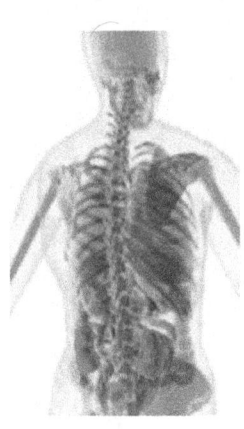

This technique relies on a stab to the kidney to induce immediate shock. The kidney is relatively accessible and by inducing shock with such a stab, the soldier has the time to cut the sentry's throat. The soldier completes his stalk and stabs the kidney by pulling the sentry's balance backward and downward and inserts the knife upward against his weight. See Step 1. The sentry will possibly gasp at this point, but shock immediately follows. By using the sentry's body weight that is falling downward and turning, the soldier executes a cut across the front of the throat. This completely severs the trachea and carotid arteries. See Step 2.

Many older military manuals teach the kidney stab as a quick, sentry kill. Some of these manuals claim that the stab is so painful and deadly that the sentry cannot even utter a sound and dies quickly. Yet, veterans and forensic experts declare the kidney stab is not so quiet and not always so successful.

Classical Military Sentry Knife Kill Scenarios

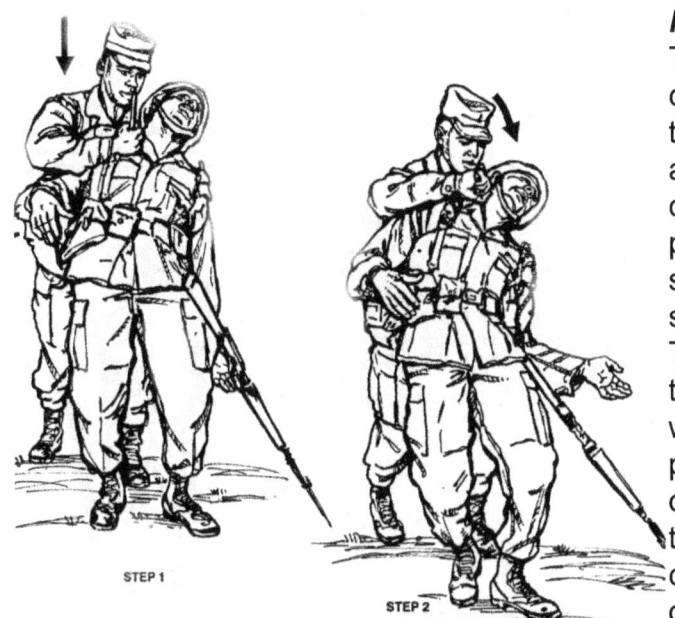

Pectoral Muscle Strike, Throat Cut
The display to the right is one of the classic problems with old-school, military knife sentry killing that veterans and the enlightened, combat instructors complain about. The soldier strikes the pectoral muscle in an effort to stun the sentry, then stabs the throat area. This stunning strike is a wasted second. That second should be spent attacking the throat. Second, consider the gear worn, even in the uniforms of military past, least of all by soldiers today. That downward, stunning, pommel strike to the brachial plexus area might well be covered by a vest full of ballistic plates or magazine pouches. A soldier should ignore the stunning strike to chest and go straight for the throat.

Nose Pinch, Mouth Grab, Throat Cut
Another classic rear-approach sentry kill. The support hand covers both the nose (somehow pinching it) as well as the mouth) Naysayers will report that, given the struggling of the sentry, the soldier may wound himself.

Crush Larynx, Subclavian Artery Stab
Crush the sentry's larynx by inserting the thumb and two or three fingers behind the larynx, then twist and crush it. The artery is stabbed. naysayers will report that, given the struggling of the sentry, the soldier's knife is too close to soldiers support hand.

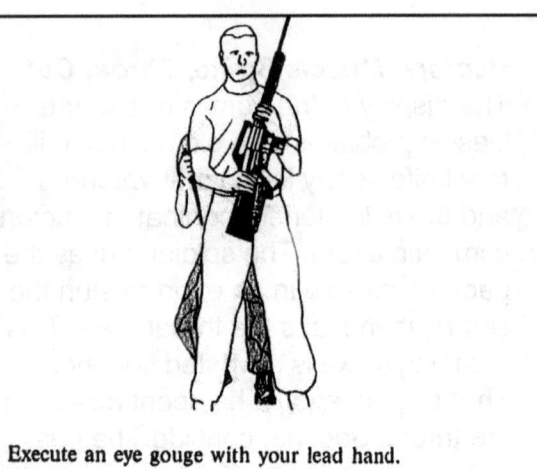

Execute an eye gouge with your lead hand.

Gouge the opponent's right eye socket and snap his head back to expose the throat.

- Plunge the blade into the left side of the opponent's throat.

- Snap the opponent's head violently to the left.
- Rip the blade across the opponent's throat to the right.

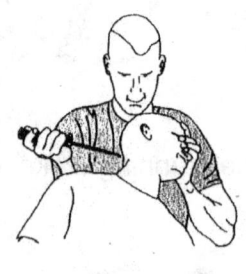

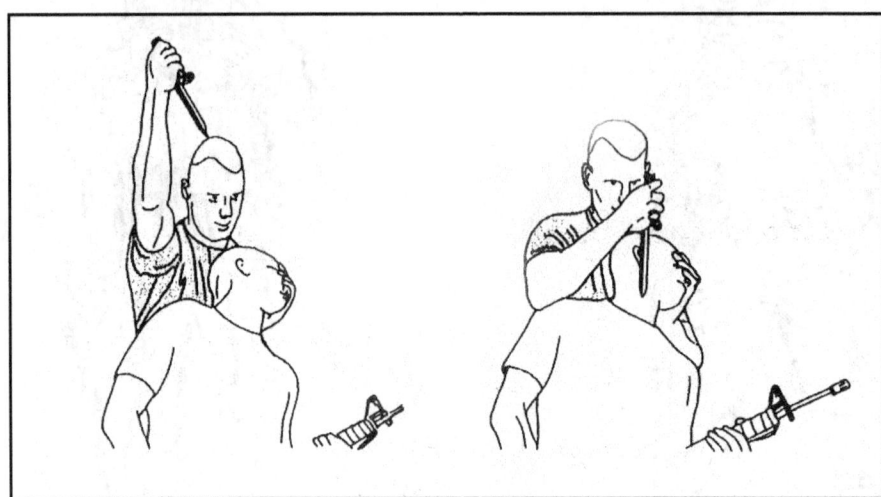

It is apparent that while any of these strategies might work, the naysayers will argue that even the most surprised sentry still may reflexively react, therefore the soldier should avoid any stabbing or slashing knife attacks that point the edges or tip of the knife back at themselves. Study the preceding drawings from actual military manuals and consider the potential for self-wounding should the attack become even the briefest of struggles.

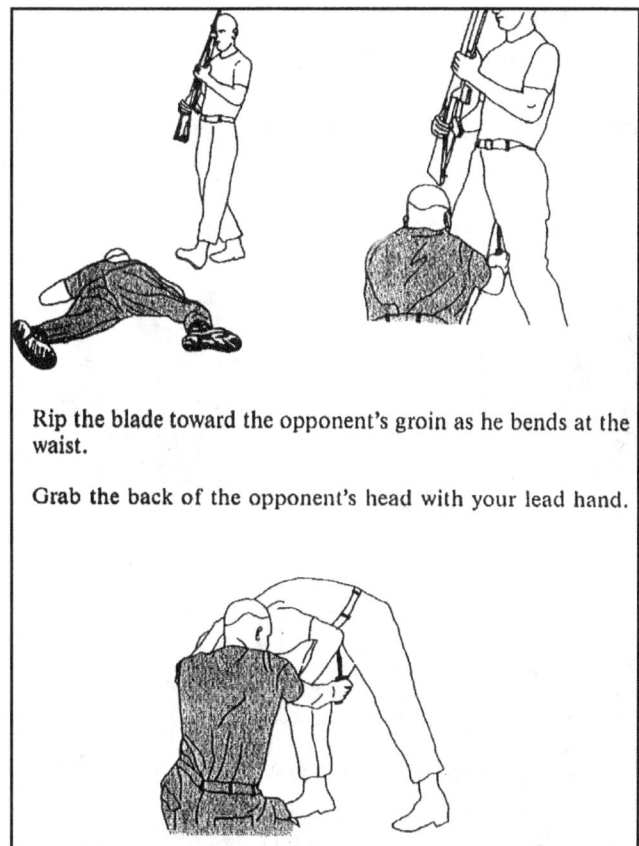

Rip the blade toward the opponent's groin as he bends at the waist.

Grab the back of the opponent's head with your lead hand.

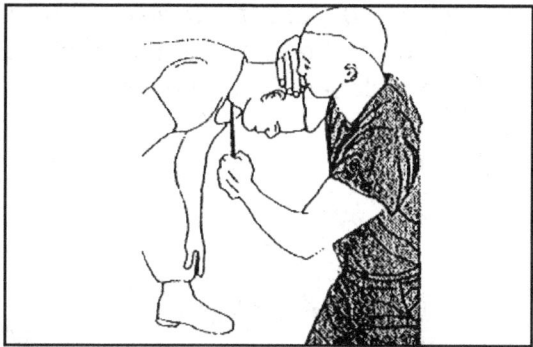

Not all sentry kills are executed from the rear. This is a U.S. Marine scenario attacking a walking guard. Stealth approach in these situations is the most difficult. In this scenario, the soldier positions himself on the walking path of the sentry. He lunges forward and stabs the groin, a wounding from which, the soldier can expect the sentry to bend over. The soldier grabs the head of the sentry and pulls him downward. Then the soldier stabs the neck of the sentry. He pulls the sentry off of his feet.

The sentry or target is not always approached from the rear. They may be approached by all angles and often in disguises. This type of disguised approach enters into the deeper subject of assassination.

Silence! The Follow-up Smothering
To further ensure silence the soldier may have to cover the mouth, throat and chest wounds of the sentry.

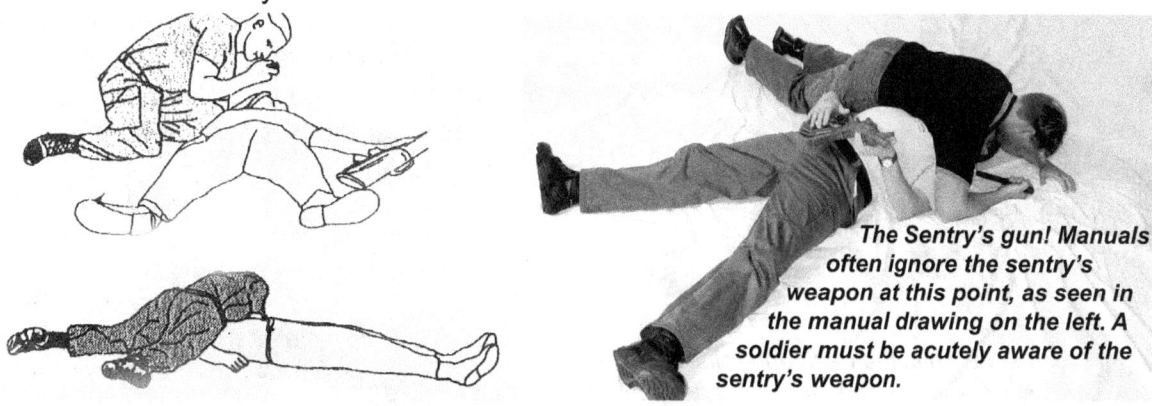

The Sentry's gun! Manuals often ignore the sentry's weapon at this point, as seen in the manual drawing on the left. A soldier must be acutely aware of the sentry's weapon.

The soldier must remember that slit and crushed windpipes may cause sucking sounds, alerting other nearby sentries. Also, the loss of breath may cause desperate, "choking" body movements that force the sentry into radical "air swimming," movements.

An Unknown Counter to Many Common Sentry Kills

This is one of the relatively unknown counters to many rear-approach, knife sentry kills that I have developed through experimentation. It is based on the knowledge that most countries teach their soldiers to capture and/or cup the mouth of their targets as the first official step in the process.

The common attack. The sentry sees the capture hand first. He...

...immediately drops. Plummets. This will confound most throat slashers and stabbers.

At times in training, the sentry has fallen between the legs of the attacker. He brings his gun to bear.

The sentry hits the deck and rolls, bringing his gun to bear on the surprised enemy.

The base of the skull, saber stab, long known as the "egg scrambler" in various military quarters is impervious to this "drop" counter because once the sentry drops down when he sees the capture hand sweeping in on him?

He would be dropping down on the point of the knife. This is but one reason why the *Egg Scrambler* is a safer method of choice in sentry elimination.

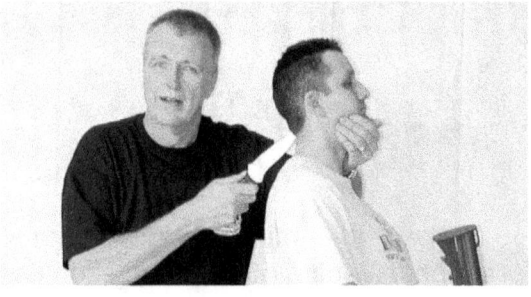

Sentry Duties

The 11 *General Orders of a Sentry* as required in U.S. Marine Corps Boot Camp as well as after, when performing sentry duty. They are universal to all services. Recruits learn these orders verbatim while at boot camp and are often expected to retain the knowledge to use for the remainder of their military careers. It is not uncommon for a drill instructor or (after boot camp) an inspecting officer to ask a question such as, "What is your sixth general order?" and expect an immediate (and correct) reply.

1: Take charge of this post and all government property in view.

2: Walk my post in a military manner, keeping always on the alert and observing everything that takes place within sight or hearing.

3: Report all violations of orders I am instructed to enforce.

4: To repeat all calls [from posts] more distant from the guardhouse than my own.

5: Quit my post only when properly relieved.

6: To receive, obey, and pass on to the sentry who relieves me, all orders from the Commanding Officer, Officer of the Day, Officers, and Non-Commissioned Officers of the guard only.

7: Talk to no one except in the line of duty.

8: Give the alarm in case of fire or disorder.

9: To call the Corporal of the Guard in any case not covered by instructions.

10: Salute all officers and all colors and standards not cased.

11: Be especially watchful at night and during the time for challenging, to challenge all persons on or near my post, and to allow no one to pass without proper authority.

Chapter 30: The Four Knife Grappling, Combat Scenario Modules

We have thus far covered a great deal of information. Philosophy. Law. Ethics. Morality. Footwork. Maneuvering, Strikes. Support. Blocking. Passing. Grappling. And, an introduction to placing these methods into sampled combat scenarios. Now we ask practitioners to freestyle situational problems in three common areas:

1: The Spartan Module – Standing, when the impact disarm works!
2: The Chain of the Knife Module – Standing, when the impact disarm didn't work.
3: Death Grip of the Knife Standing, when you grab him and he grabs you.
4: Rattlesnake! – Grounded, when you knife fight on the ground in the Spartan, Chain and Death Grips.

I am also a proponent of my *Stop 6* program, which is exercising in the 6 common collisions in any fight. What we are about to cover here already covers the *Stop 6* progression, just in a different way. Remember all our pre-fight, pre-crime advice and the knife quick draws in this book. These modules occur *after* all that. First we will start with the Spartan Module.

Module 1: The Spartan Impact Disarm Combat Scenarios

I have called this module after the Spartans since the 1990s, way before the popular movies captured the imaginations of people. When the Spartans and other legendary soldiers of the past such as centurions and legionnaires could not lunge out and stab the torsos of their enemies, they would hack at the closer weapon-bearing limbs. This is one early historical strategy, found in many times and places.

The disarm may not be achieved, but in this Spartan model, the successful, impact weapon disarm is explored, developed and emphasized. Called the Spartan exercises and scenarios, the soldier strikes the weapon-bearing limb of the enemy, and the impact is strong enough to dislodge the weapon from the enemy's grip. Next, the soldier must fight on from there, taking appropriate action to conclude the situational encounter.

Spartan Briefing 1: The Enemy Weapons.
The enemy may be armed with any type of weapon, hand, stick, knife or gun. If the defender strikes the weapon-bearing limb of an attacking enemy with his knife, the enemy's limb may be diminished in strength and performance and at times drop the weapon. The defender may also strike any part of the body which might cause a shocking, physical disruption of the hand grip, causing the weapon to drop.

The enemy may attack with:
  Enemy Weapon 1: Any attack from the non-weapon arm, including the shoulder ram, elbow, forearm or hand.

  Enemy Weapon 2: Any kicking attack.

  Enemy Weapon 3: Any impact weapon attack.

  Enemy Weapon 4: Any edged-weapon attack.

  Enemy Weapon 5: Any firearm. The attack begins when the soldier sees the weapon being carried, being pulled, pulled and used as a threat, or being used.

Spartan! The weapon-bearing limb is struck.

Spartan Briefing 2: Type of Spartan Disarming Strikes.
Any solid strike to the enemy's body might cause an impact disarm. The main Spartan disarming strikes are:

  Strike Concept 1: Knife strike tools:
    - a saber grip, or a reverse grip.
      * slashing strike.
      * stabbing strike.
      * a power block/hack.
      * pommel strike.

  Strike Concept 2: Knife strike to:
    - weapon limb.
    - other in-range body targets.

  Strike Concept 3: Support strikes
    - all support limb strikes.
    - all kicks.

  Strike Concept 4: Be prepared that the strike may not disarm. The defender must fight on with other skills developed in subsequent sections of the book. In this isolated study, the impact actually worked.

Spartan Briefing 3: The Three Spartan Encounters
The impact disarm strike might occur in three possible events. The first is the pre-emptive seconds during the enemy's initial weapon quick draw. In the case of a long gun on a sling this means the enemy is grabbing his weapon and bringing it up for use. The second is when the weapon is presented as a threat. The third is when the enemy attacks. Types of Spartan impact disarm encounters are:

Spartan Encounter 1: Disarm during the weapon quick draw.
Spartan Encounter 2: Disarm during the weapon presentation when the weapon is posed as a threat.

Spartan Encounter 3: Disarm during an attack.

- Impacts with knife attacks.
    - stab or slash.
    - thrust or jab.
    - on the Combat Clock.

- Stick attacks.
    - stab or slash.
    - thrust or jab.
    - on the Combat Clock.

- Rifle bayonet attacks
    - stab or slash.
    - thrust or jab.
    - on the Combat Clock.

The defender interrupts the enemy quick draw as he tries to pull and open a tactical folder –always a flimsy, weak grip.

- Firearm threats/attacks.
    - If the weapon is actually firing, the odds are possible but slim. It is situational. Best to counter this situation in the quick draw and presentation phases.

An incoming baton attack. Knife hits anywhere on the weapon-bearing limb.

Spartan Encounter 1: Knife attack during the weapon quick draw
No matter the enemy's weapon, the soldier trains to knife strike the quick draw process. The soldier sees the draw, strikes the limb, observes a disarm, and finishes the fight as needed based on the mission with threats, or more knife attacks and/or unarmed support skills and a takedown. No matter the enemy's weapon being drawn, the defender trains to knife strike the quick draw process in this part of the Spartan Training Module.

The defender:
  Step 1: sees through any fakes or subterfuge and sees the draw.

  Step 2: knife strikes the limb.

  Step 3: observes a disarm.

  Step 4: follows-up with needed action.
    – threats.
    – more attacks.

The defender lunges forward to strike the weapon limb and interrupt the enemy's weapon quick draw.

Practice the Spartan Knife Counter to Quick Draw Scenarios.
In each set, the soldier lunges forward to interrupt a weapon quick draw.

Spartan Counter QD Scenarios Set 1: The trainer draws a fixed blade knife:
  – from a right side belt carry.
  – from a left side belt carry.
  – from cross draw shoulder sheath carry.
  – from other possible body sites.
  – from nearby off-body sites.

Spartan Counter QD Scenarios Set 2: The trainer draws and tries to pull and open a folding knife:
  – from a right side pocket or belt.
  – from a left side pocket or belt.
  – from other possible body sites.
  – from nearby off-body sites.

Spartan Counter QD Scenarios Set 3: The trainer draws an expandable baton:
  – from a right side belt or pocket.
  – from a left side belt or pocket.
  – from a cross draw shoulder sheath.
  – from other possible body sites.
  – from nearby off-body sites.

Spartan Counter QD Scenarios Set 4: The trainer draws a pistol from a holster:
- on right side belt.
- on left side belt.
- from cross draw shoulder holster.
- from other possible body sites.

Spartan Counter QD Scenarios Set 5: The trainer has a long gun.
- hanging on a sling.
- nearby and lunges for it.

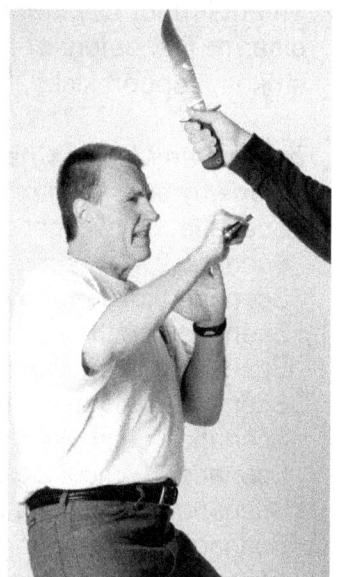

Spartan Encounter 2: Knife attack during a weapon attack
The enemy attacks the defender with a empty hands, a stick and a knife in thrusting and hooking deliveries. The following combat clock format develops the Spartan response in skill development drills. The Spartan Skill Developing Drills training progression is as follows:

Development Set 1: The enemy attacks with a stick or any impact weapon. The defender stabs or slashes the weapon-bearing in an attempt to block the incoming attack.

A thrusting 12 o'clock.
A thrusting 3 o'clock.
A thrusting 6 o'clock.   Saber grip.
A thrusting 9 o'clock.   ><
A hooking 12 o'clock.   Reverse grip.
A hooking 3 o'clock.
A hooking 6 o'clock.
A hooking 9 o'clock.

Development Set 2: The enemy attacks with a stick or any impact weapon. The defender stabs or slashes the weapon-bearing arm in an attempt to block the incoming attack.

A thrusting 12 o'clock.
A thrusting 3 o'clock.
A thrusting 6 o'clock.   Saber grip.
A thrusting 9 o'clock.   ><
A hooking 12 o'clock.   Reverse pick grip.
A hooking 3 o'clock.
A hooking 6 o'clock.
A hooking 9 o'clock.

Development Set 3: Repeat the above, and the soldier adds a cautionary strike to the enemy's support, empty hand should the hand appears ready to attack.

The Spartan Knife Combat Scenarios
The defender hits the weapon-bearing limb. In this segment, the impact causes a disarm. The defender moves in to kill or capture the enemy with a series of support strikes, support kicks, knife attacks, a takedown and a finish appropriate for the situation.

The Spartan Module Chart
The enemy attacks with:
  Scenario 1: A weapon quick draw.
  Scenario 2: A weapon threat presentation.
  Scenario 3: A thrusting 12 o'clock attack.
  Scenario 4: A thrusting 3 o'clock attack.
  Scenario 5: A thrusting 6 o'clock attack.
  Scenario 6: A thrusting 9 o'clock attack.
  Scenario 7: A hooking 12 o'clock attack. ><
  Scenario 8: A hooking 3 o'clock attack.
  Scenario 9: A hooking 6 o'clock attack.
  Scenario 10: A hooking 9 o'clock attack.
Exercise some combat scenarios with less-than-lethal applications of the knife.

Spartan Defender responds with:
  – Uses footwork.
  – Sees and seeks a weapon disarm or,
  – Strikes the quick draw or,
  – Strikes the weapon bearing limb.
  – Support hand strikes if can be done.
  – Support kicks if can be done.
  – Monitor his support hand. Hit it with knife.
  – Blocks enemy support hand strikes.
  – Decides if disarms enemy is "through?"
  – Does a takedown if needed.
  – Does not let the enemy recover the dropped weapon.

Spartan Sample: The 12 o'clock stick attack

The high attack.

The Spartan strike. A disarm!

The support hand strike.

Clavicle notch takedown.

Spartan Sample: The 9 o'clock stick attack

The side attack.   The knife blocks/strikes for a disarm.

A throat slash starts the finish, onto a take down.

Spartan Sample: the 6 o'clock attack

The low attack. The Spartan impact weapon disarm strike. Monitor the support hand. Pelvis, groin stab, push take down.

Spartan Combat Scenario Sample: The 3 o'clock knife attack

The enemy attacks at 3 o'clock.

The defender strikes. A disarm!

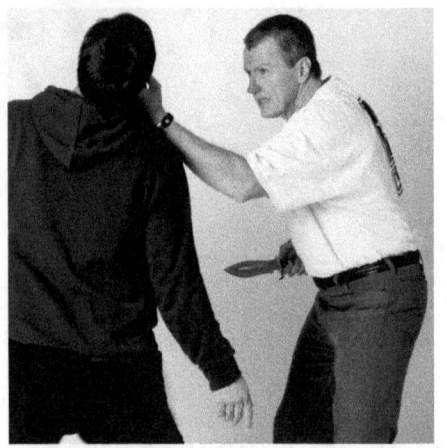

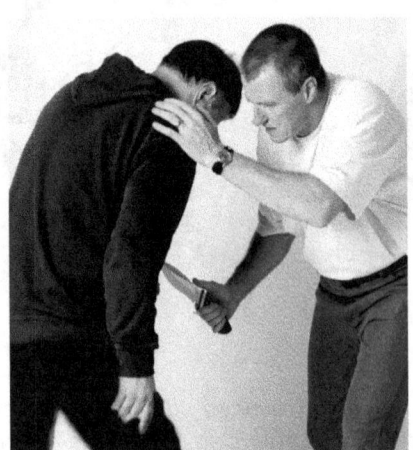

The defender strikes and strikes. The enemy bends over from a low stab and is hammered by hammer-fist/forearm or/and knife pommels to the ground.

The Spartan Dilemma! The Dropped Knife

The impact disarm worked! The knife (or gun, or whatever weapon) is on the ground or floor. What happens next? An inexperienced, virgin observer, (like a juror?) might consider that the death threat, the lethal force against you is over. After all, he dropped the knife after you disarmed him. But that knife is right there within a torso bend or a body dive for a recovery in a second.

Is his deadly intent still there in his mind. Is this death threat over? His spirit broken? Unlikely. Are you a mind reader who can tell? No you aren't. You can't.

You must do what you can to keep him away from the weapon; and if that means, under the circumstances, deadly force to save your life or the lives of others – then so be it.

Perhaps he is dazed from your response? This is the time when all your less than lethal combatives training may come into play. Even if you are holding a knife you can deliver less than lethal applications with it in your hand, as displayed earlier in this book.

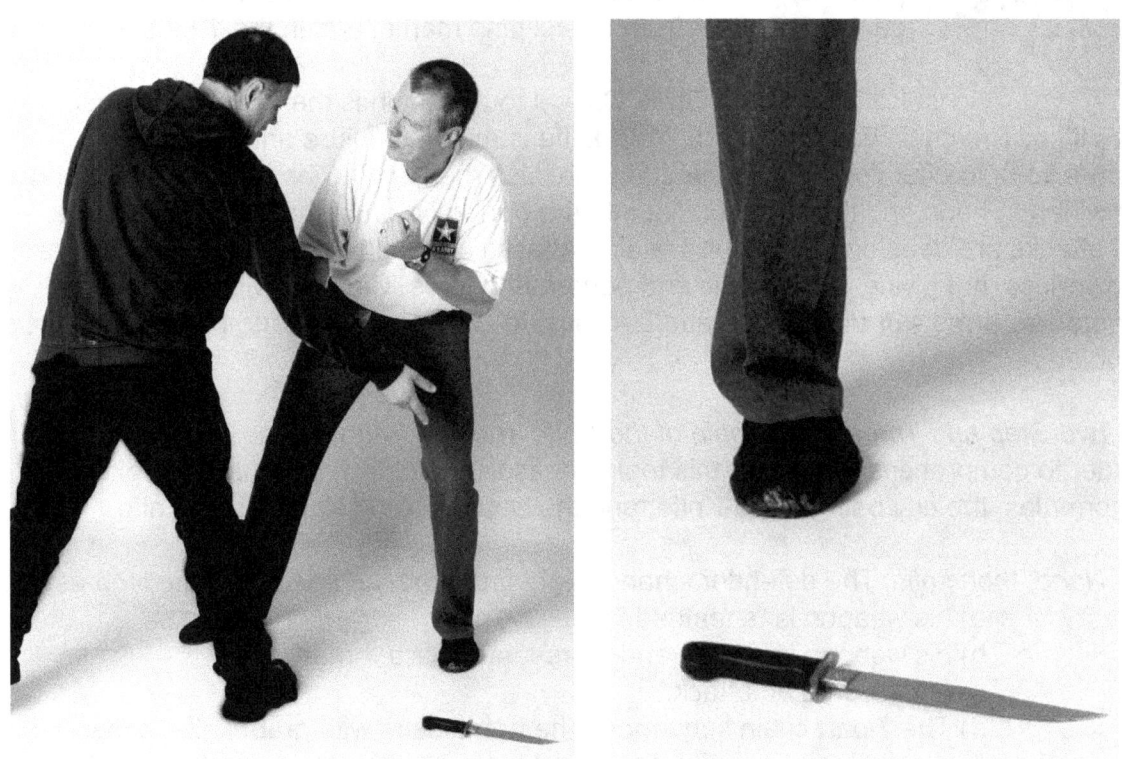

The dropped knife is still a deadly situation. He can grab it in a second!

Module 2: The Chain of the Knife, "Grab" Combat Scenarios

The Chain of the knife is about the grab. At the core of the knife chain drills, is the word *chain*. The defender's free, support hand grabs the enemy's weapon-bearing limb in the first, or the second, or later steps/events of the fight. The grab acts as the first or second link in the chain of survival. This next knife, grappling combat scenario series is based on when the defender grabs the enemy's weapon-bearing limb. It appears in a progression here for when the previous Spartan Drill Impact Disarm didn't work. Or, the weapon limb grab might be before such a disarming strike.

If an impact disarm fails, no disarm is achieved! One smart option next is to secure a "link," a grab on the weapon bearing limb. The first reference to "knife" in the sequence of "knife-grab-knife." Your blade, whether saber or reverse, hits the weapon, bearing limb. You grab the limb. You invade in. I first learned of this nickname and organized concept from the Filipino martial arts in the 1980s.

The Critical Grab. Critical to the chain is the grab of the opponent's limb. The chain drill is the grab! Your knife is meant to injure and freeze the enemy's limb, to offer you the chance at a grab. Use the **Big C Clamp** approach. Open your widened hand and then thrust it out into the general path of the incoming attack. Real attacks are deep and powerful. Training attacks are usually short and half powered, or powerless, not giving you realistic and penetrating force. Clothing will interfere with your grab as arms can slip forward and back inside sleeves. Sweat and blood will also be a problem.

The Two-Step and Three-Step Chain of the Knife Training Sequences:
In order to comprehensively cover this topic, the isolation drill study must be broken into two formulas, the first being Hand-Knife, and the second being Knife-Hand-Knife.

Hand, then knife: The defender's hand might first grab the enemy's knife limb as:
- a) his weapon is "sheathed."
- b) his weapon is drawn, and it presented for a threat.
- c) or delivered an attack.
- d) The 2-part chain sequence. The hand leads with grab, knife follows.
  - Hand First: The free hand grabs the weapon limb, the first link in the chain of survival.
  - Knife Next: The knife continues with the attack, staying aware of the enemy's other hand. If that other hand offers no immediate threat? Then move on the vital attack.

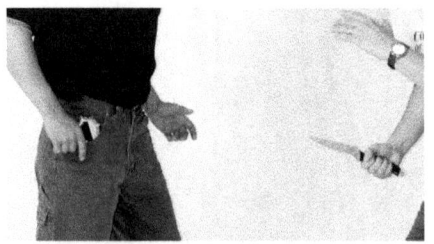 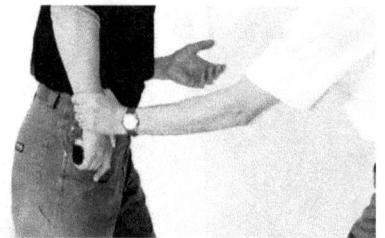

Knife-Hand-Knife, The 3-Part Chain Sequence. The knife leads, the hand grabs, the knife attacks

    a) the weapon is drawn.
    b) as it is presented for a threat.
    c) or delivered as an attack.
    d) the 3-part sequence.
        Knife First: The knife makes contact with the weapon-bearing limb.
        Hand Next: The free hand grabs the knife limb, the first link in the chain of survival.
        Knife Next: The knifer continues with the attack, aware of the enemy's other hand. If that other hand offers no immediate threat? Then move on the vital attack. Or the defender's knife has hit the enemy's weapon-bearing limb, possibly injuring it, slowing it down, or even stopping it.

Hand-Knife! The Chain of the Knife Drill 1: Grab and Knife Attack

This is a preparatory skill drill. In these studies, the trainer draws a weapon. The trainer presents the weapon. The trainer attacks on the four corners of the Combat Clock. The trainee sees that the weapon-bearing limb can be stopped and caught – usually at the forearm, hopefully at the wrist – and does so. Should the trainer's other empty hand not appear threatening, the trainee can forego striking that hand and attack vital targets.

Grab First, Basic Set 1: The trainer attempts to draw a pistol or a knife from a carry site. The trainee grabs the drawing arm, checks or strikes the other arm, then attacks the trainer at a vital target.

 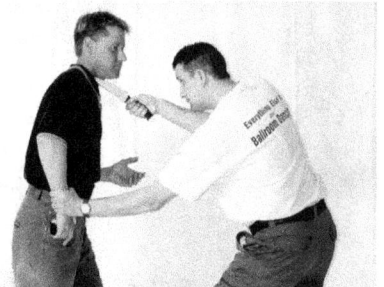

In two beats. The trainer grabs for a pistol, or any weapon. The trainee grabs the weapon-bearing limb and stabs a vital target.

Grab First, Basic Set 2: The trainer has the weapon drawn and presented. This is usually at the axis of the Combat Clock. The trainee grabs the weapon-bearing limb, checks or strikes the other arm, then attacks the trainer at a vital target.

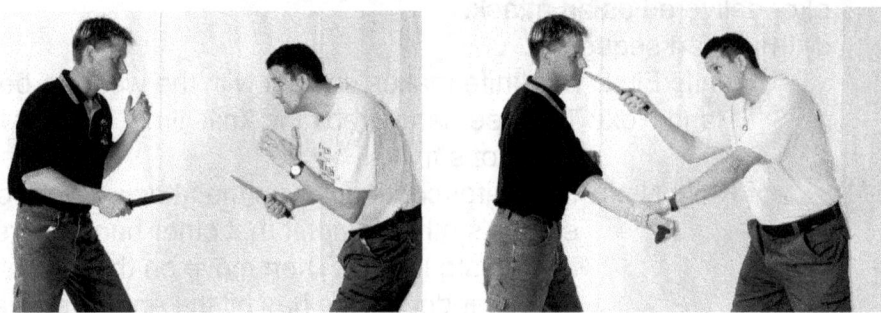

The Trainer threatens. The trainee gets a grab on the weapon limb and strikes.

Grab First, Basic Set 3: The trainer attacks at 12 o'clock high. The trainee seizes the trainer's weapon bearing limb, checks or strikes the trainer's support limb, then knife attacks the trainer.

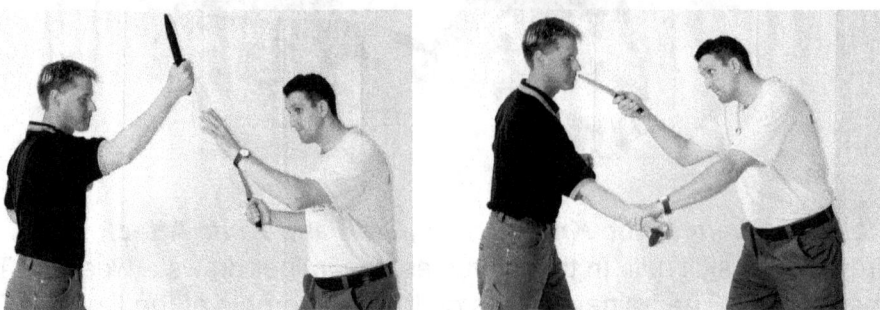

The Trainer threatens. The trainee gets a grab on the weapon limb and strikes.

Grab First, Basic Set 4: The trainer attacks at 3 o'clock. The trainee seizes the trainer's weapon bearing limb, checks or strikes the trainer's support limb, then knife attacks the trainer.

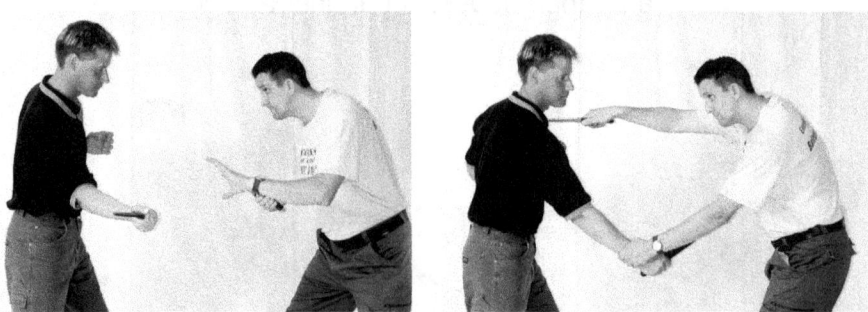

The right-side slash comes in. The Trainee grabs the limb and attacks.

Grab First, Basic Set 5: The trainer attacks at 6 o'clock low. The trainee seizes the trainer's weapon bearing limb, checks or strikes the trainer's support limb, then knife attacks the trainer.

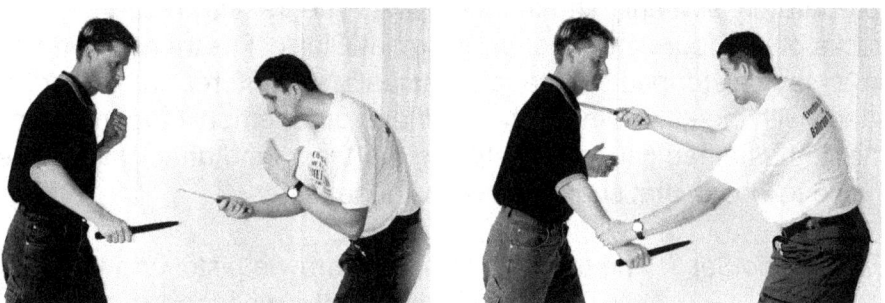

A low attack comes in. The trainee grabs and attacks a vital target.

Grab First, Basic Set 6: The trainer attacks at 9 o'clock. The trainee seizes the trainer's weapon. bearing limb, checks or strikes the trainer's support limb, then knife attacks the trainer.

A backhand attack comes in. The trainee grabs and attacks a vital target.

Grab First, Advanced Set 1. The trainer attacks on all the 12 Combat Clock angles.

Grab First, Advanced Set 2: The trainer's strikes and the trainer's kicks. After a defender grabs the weapon limb, the defenders next biggest problem is the trainer's support limb strike, then second, the kick. Most people are not kickers, but you must consider the possibility. The trainer, once grabbed adds a:
- hooking strike to the aforementioned steps, which must be blocked by the trainee.
- thrusting strike to the aforementioned steps, which must be blocked by the trainee.
- a kick.

Knife-Hand-Knife! The Chain of the Knife Drill 2: The Knife Strike, the Grab and Knife Attack Sequence

This is a preparatory skill drill. The trainer draws. The trainer presents the weapon. The trainer attacks on the four corners of the Combat Clock. The trainee sees that the weapon-bearing limb may be stopped and caught – usually at the forearm – and does so, after hitting the attack with the knife. Should the trainer's other empty hand not appear threatening at that instant, the trainee can forego striking that hand and attack vital targets, otherwise, it would be wise to whack that support hand with the knife.

Knife First, Basic Set 1: The trainer attempts to draw a pistol or a knife from a carry site. The trainee strikes the weapon-bearing limb. The trainee grabs the weapon-bearing limb, hits his other support limb if it is threatening. Then he attacks the trainer at a vital target.

The trainer touches his weapon. The trainee strikes the limb, grabs the limb and attacks.

Knife First, Basic Set 2: The trainer has the weapon drawn and presented. This is usually at the axis of the clock. The trainee strikes the weapon-bearing limb with his knife, then grabs the weapon-bearing limb, checks or strikes the trainer's other arm if a threat, then attacks the trainer at a vital target.

The trainer presents his weapon. The trainee hits the limb, grabs the limb, hits the other hand if needed, then attacks vital targets.

Knife First, Basic Set 3: The trainer attacks at 12 o'clock high. The trainee knife strikes the weapon bearing limb, the free hand seizes the trainer's weapon bearing limb, checks or strikes the trainer's support limb, then knife attacks the trainer.

The trainer presents his weapon. The trainee hits the limb, grabs the limb, hits the other hand if needed, then attacks vital targets.

Knife First, Basic Set 4: The trainer attacks at his 3 o'clock. The trainee knife strikes he weapon-bearing limb, the free hand seizes the trainer's weapon bearing limb. The trainee checks or strikes the trainer's support limb if need be, then knife attacks the trainer.

The trainer attacks at 3 o'clock. The trainee hits the limb, grabs the limb, hits the other hand if needed, then attacks vital targets.

Knife First, Basic Set 5: The trainer attacks at 6 low. The trainee knife strikes the weapon bearing limb, the free hand seizes the trainer's weapon bearing limb, checks or strikes the trainer's support limb, then knife attacks the trainer.

The trainer attacks at 6 o'clock low. The trainee hits the limb, grabs the limb, hits the other hand if needed, then attacks vital targets.

Knife First, Basic Set 6: The trainer attacks at his 9 o'clock high. The trainee knife strikes the weapon bearing limb, then the free hand seizes the trainer's weapon bearing limb. He checks or strikes the trainer's support limb, then knife attacks the trainer.

Knife First, Advanced Set 1. The trainer attacks on all the 12 Combat Clock angles.

Knife First, Advanced Set 2: The trainer's strikes and the trainer's kicks. After a defender grabs the weapon limb, the defenders next biggest problem is the trainer's support limb strike, then second, the kick. Most people are not kickers, but you must consider the possibility. The trainer, once grabbed adds a:
- hooking strike to the aforementioned steps, which must be blocked by the trainee.
- thrusting strike to the aforementioned steps, which must be blocked by the trainee.
- a kick.

Grab First, or Knife First, Chain Series Interrupted by punches, Set 1 – The Hook
In this sample, the trainer attacks at 3 o'clock. The trainer is in a position to strike the incoming limb with his knife. The trainee executes this block/strike and the trainer immediately throws a hooking strike. The trainee hits the incoming hooking arm. He stops it and pushes it back to "disconnect" the mission of the strike. Once the arm is stopped and pushed back, next the trainee strikes a vital target. Using the four-corners of the Combat Clock, practice blocking the trainer's half-beat strike.

Grab First, or Knife First, Chain Series Interrupted by punches, Set 2 – The Thrust
In this sample, the trainer attacks at 3 o'clock. The trainer is in a position to strike the incoming limb with his knife. The trainee executes this block/strike and the trainer immediately throws a straight strike. The trainee dodges and hits/passes the incoming strike. Next the trainee strikes a vital target. Using the four-corners of the Combat Clock, practice blocking the trainer's half-beat strike.

Grab First, Advanced Set 1. The trainer attacks on all the 12 Combat Clock angles.

Grab First, Advanced Set 2: The trainer's strikes and the trainer's kicks. After a defender grabs the weapon limb, the defenders next biggest problem is the trainer's support limb strike, then second, the kick. Most people are not kickers, but you must consider the possibility. The trainer, once grabbed adds a:

- hooking strike to the aforementioned steps, which must be blocked by the trainee.
- thrusting strike to the aforementioned steps, which must be blocked by the trainee.
- a kick.

The Chain of the Knife Combat Scenarios

The defender hits the weapon-bearing limb. In this segment, the impact causes a disarm. The defender moves in to kill or capture the enemy with a series of support strikes, support kicks, knife attacks, a takedown and a finish appropriate for the situation.

The Chain of the Knife Module Chart
The enemy attacks with:
- Scenario 1: A weapon quick draw.
- Scenario 2: A weapon threat presentation.
- Scenario 3: A thrusting 12 o'clock attack.
- Scenario 4: A thrusting 3 o'clock attack.
- Scenario 5: A thrusting 6 o'clock attack.
- Scenario 6: A thrusting 9 o'clock attack.
- Scenario 7: A hooking 12 o'clock attack.
- Scenario 8: A hooking 3 o'clock attack.
- Scenario 9: A hooking 6 o'clock attack.
- Scenario 10: A hooking 9 o'clock attack.

Exercise some combat scenarios with less-than-lethal applications of the knife.

Chain Defender responds with:
- Uses footwork.
- Sees and seeks a weapon disarm or,
- has to grab the weapon-bearing limb.
- Strikes the quick draw or,
- Strikes the weapon bearing limb.
- Support hand strikes if needed.
- Support kicks if can be done.
- Monitor his support hand. Hit it with knife.
- Blocks enemy support hand strikes.
- Does a takedown if needed.
- Does not let the enemy recover the dropped weapon.

Module 3: The Death Grip of the Knife Combat Scenarios

In the "Death Grip of the Knife" I emphasizes combat scenarios when both fighters are grabbing each other. The classic "knife death grip" is when the enemy holds the defender's weapon-bearing limb and the defender in turn, does the same with the enemy's weapon limb. It is the natural training progression to follow the "Chain of the Knife." Dissecting the possibilities catch positions creates a battle plan for the combat. The enemy may be brandishing an edges weapon, an impact weapon, even a handgun. Knives held may be in a saber or reverse grip.

The 5 Common Knife Death Grip Catches:
Clutches 1: High in the death grips of.
Clutches 2: Low in the death grips of.
Clutches 3: Split high and low in the death grips of.
Clutches 4: Righty vs. lefty grips in the death grips of.
Clutches 5: Arm wraps and near bear hugs in the death grips of.

High. *Low.* *Split .*

Righty versus lefty. *Arm wraps and near bear hugs.*

The "accordion." Given the in-and-out motion of a fight, sometimes chests "bump."

Being "caught red handed" can occur on the ground and floors too. We will cover those problems.

Saber and reverse grips many vary. There have been some law enforcement studies that suggest high or higher attacks often involve the reverse grip and mid-to-lower attacks often are saber grips. These are natural. However there are a few newer knife courses that are making practitioners pick one or the other for everything, which can be limiting.

3 stabs/cuts to his defense.

Defense weakened and 3 to his offense.

The same for the low line catches. You stab the same side or cross over and stand the other the forearm.

High catch basics
    – 3 to the catching limb (his defense).
    – 3 to the weapon limb (his offense).

Low catch basics
    – 3 to the catching limb (his defense).
    – 3 to the weapon limb (his offense).

Mixed catches? Mix these basic moves and experiment. For example, you might kick the opponent, spit in his eye and attack the other limb, wounding his offense.

Mixed? Those 12 would be basic training. But what if the grips weren't just like these? Perfect high and perfect low? Like a saber grip high and not a reverse grip when the knife tip is not near his forearm? What if when low and with a reverse grip, and that tip isn't near the forearm? What if one grip was high? The other low? Mix up the responses.

The Knife Death Grip, Statue Kick Drill

You can get in the Death Grip situations, usually mixed or high grips, you might get a quick kick in. To practice this, we use the statue drill, and the trainer stands, legs apart so the trainee can experiment and fabularize his or herself with the progression of outside-inside-inside-outside the trainers legs. The trainer, being the statue, allows the trainee to experiment.

The nature of the arms distance, the *Stop 2* distance, allows for some freedom in kicking without being so easily pushed off balance, as one might with an arm wrap.

The fact that the enemy's hands/forearms are being held, hinders the opponent's ability to stab your kicks. If the hands are low? This availability might lead to a kicking leg slash or stab.

The *Force Necessary: Hand* course covers these kicks one by one in it's progression. Look there for detailed instruction if you need it. But here, do your best, as none of these kicks are "rocket science." This Statue Drill exercise emphasizes the:

Frontal snapping kick to shin and groin.

* Frontal snap kick.
   – left kick to shin – left to groin – right to groin – right to shin. Next work back from right to left.

* Frontal shin kick with shoe toes or side of the shoe.
   – left to right shin – right to left shin – left to left shin right to left shin. Next work back from right to left.

* Stomp kick.
   – left to right foot top – right to left foot top – left to left foot – right to left foot top. Next work back from right to left.

Frontal snapping kick to shin and groin.

* Thrust kick.
   – left to right knee – right to left knee – left to left knee – right to left knee. Next work back from right to left.

* Round kick.
   – left to right leg – right to left leg – left to left leg – right to left leg. Next work back from right to left.

* Knees (and there are many options.
   – left to outer right – left to center right – left to groin – left to inside left – right to outer left – right to center left – right to groin – right to inner right. (there are more – see FN: Hand Level 3. Next work back from right to left.

* Whatever other kicks you think you can get in.

Round kick to legs.

A Study of the Major Releases via the Death Grip
Your knife lower limb/hand gets grabbed. High or low. These work versus one hand only grabs too and usually in Stop 2 to 6 also. This book covers *Stop 2* only, but you will be asked to review this releasing foundation as the *Stop 6* and *Training Mission* books continue.

Sample 1: The yank-out. Like a hit and retract, use your arm and whole body if need be to get put of this *Stop 2* grab. Same on the low end.

1: The yank-out release.

Sample 2: The circular release and the joint lock position releases. Rotating your knife hand clockwise or counter-clockwise can often get a release. It will depend on the grip, high or low. Sometimes a quarter-circle or a half-circle will allow you to use a quicker, joint lock release. At times, a second after the release, you can push the hand away with your saber or reverse grip, freeing a better path to your next attack, else the opponent will simply re-grab your limb. Look for this push-out opportunity in your practice.

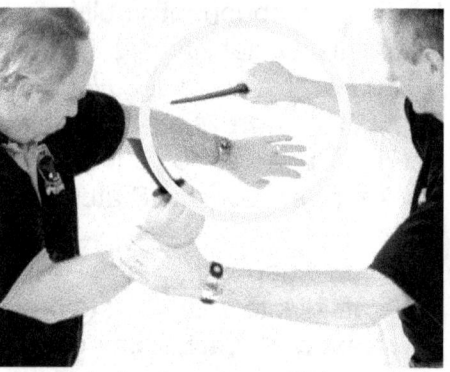
2: A circular release. This one counter-clockwise.

Sample 3: The elbow rollover. This is really a joint lock release also, but it serves an honorable mention for its easy success. If you can, it's 1) elbow up. 2) elbow over his forearm. 3) elbow down like a flapping motion. You can accompany that downward flap with a bit of a body drop against very strong grips.

Sample 4: The shoulder release. If you can't get a release from a strong grip? And you can judge from prior practice versus various heights, pull his gripping hand out from his body and get your upper torso under his arm, a bit sideways. Hammer your caught hand down as you stand up like a squat. This should get a release. Take care not to get caught in a head lock, by going a bit deeper in and a bit of sideways.

3: A joint lock release. He's in a center lock position. Jam knife down.

Sample 5: A knee release. On the low side of the clock, if you can't get a release, and you can judge from prior practice versus various heights, lift and drive you knee against his grabbing forearm, and pull your caught limb back and out.

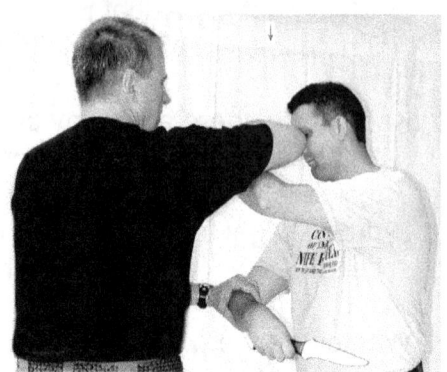
3b: An elbow rollover.

Sample: "Can't get a release?" The low arm swing

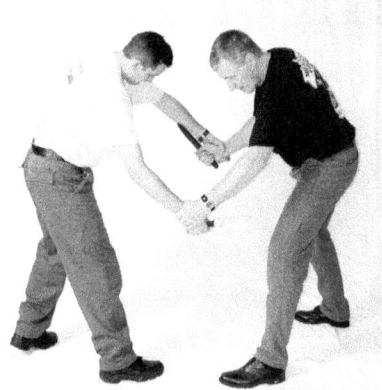

Low clutch grab.

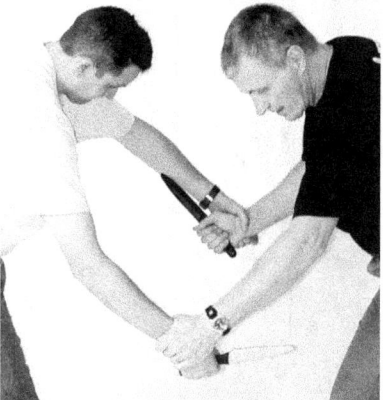

Defender's knife attacks on limb...then attacks enemy knife limb.

Defender rides the arm down into the thigh or slashes femoral artery.

The enemy bends. The soldier stabs into the jaw line. Twist the knife...

...until the blade lays across the neck. The soldier begins the rear leg sweep takedown.

Pull and leg sweep. Knife pushes against the throat.

There are many options for the low clutch grab situation. Circular releases. Pull-outs. This grab is unique because the soldier can ride the enemy's arm with a swinging motion down into his legs This may cause the enemy to step back and take him off balance, with his head forward and exposed. Whatever the follow-ups, once the enemy is grounded, the defender continues to fight on in accordance with his mission and assignment.

Sample of a release from an arm wrap
I learned this in the Philippines and it involves a hand grip switch from saber to reverse.

Straighten out the saber grip, get proper leg position and try to pull out. Cut along the way.

A common arm wrap capture, get ready to switch to a reverse grip.

Angle the reverse grip tip in, get proper leg position and pull back for a stab to the back.

The Death Grip of the Knife Workout List:
Review all prior grabbing/catching skills previously documented in
this manual. These sets isolate and skill develop a small part of the knife fight.

1) High Catch Set Standing
   - The enemy attacks. The soldier catches.
   - The soldier attacks. The enemy catches.
   - Trainee stabs defensive arm 3 times.
   - Then stabs attacking arm 3 times.
   - Experiment with slash on the attacking arm.
   - Experiment with a sudden release of a grabbed
     limb when the limb is stabbed.

2) High Catch Set Grounded
   - The enemy attacks. The soldier catches.
   - The soldier attacks. The enemy catches.
   - Trainee stabs defensive arm 3 times.
   - Then stabs attacking arm 3 times.
   - Experiment with slash on the attacking arm.
   - Experiment with a sudden release of a grabbed
     limb when the limb is stabbed.

3) Low Catch Set Standing
   - The enemy attacks. The soldier catches.
   - The soldier attacks. The enemy catches.
   - Trainee stabs defensive arm 3 times.
   - Then stabs attacking arm 3 times.
   - Experiment with slash on the attacking arm.
   - Experiment with a sudden release of a grabbed
     limb when the limb is stabbed.

4) Low Catch Set Grounded
   - The enemy attacks. The soldier catches.
   - The soldier attacks. The enemy catches.
   - Trainee stabs defensive arm 3 times.
   - Then stabs attacking arm 3 times.
   - Experiment with slash on the attacking arm.
   - Experiment with a sudden release of a grabbed
     limb when the limb is stabbed.
   *(knife ground fighting follows in the
   Rattlesnake Chapter)*

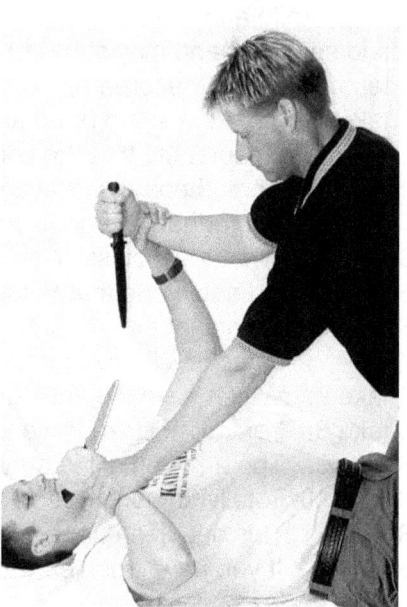

Circular Special! The Snake Disarm!

The reverse grip attack is blocked. This works best when his main or single edge is out at you.

The defender raises his elbow and forearm up to knife hand. Note the sharp bend in the forearm.

Circle the wrist. NEVER let the thumb or palm touch the enemy's forearm. This slows the circle.

Continue the circle until the knife is forced out. A soldier might slash the forearm to assist the process, or if close enough and the arm is free, even stab the enemy. The demonstrators here have stepped back to allow a clear photo view of the snaking steps. The soldier attacks for a finish.

Mandatory snake disarm knowledge and steps
This works against an ice pick attack. The circular snake disarm has been a natural occurrence in recorded knife and mixed weapon fights. So much so, that it bears recording and instructing. The circling/snaking hand may start from a common grab, or from a common forearm block.

- Snake 1: Raise the forearm
- Snake 2: Create a sharp bend at the elbow.
- Snake 3: Understand that the contact will not cause a severe cut to the blocking limb. At very worst there might be a minor cut, but this movement has been done hundreds of times with edged weapons, without any cuts.
- Snake 4: Wrist-to-wrist, start the circle.
- Snake 5: Do not let the thumb touch on the forearm.
- Snake 6: Do not let the forearms touch, the more you remain wrist to wrist, the better chance of faster, clean disarm. Keep the circle wrist-to-wrist. Keep the arm bent at the wrist.
- Snake 7: Watch were the knife drops. The tip might fall into your leg or foot. This has happened.
- Snake 8: You might strike the enemy during the process.
- Snake 9: The arm contact is against the flat of the blade.
- Snake 10: Totally untrained people have successfully snaked disarmed knife attackers.
- Snake 11: This has been done in knife ground fights.
- Snake 12: If you have troubles with this, get with a personal instructor. This is not hard to learn.

Module 4: The Rattlesnake! Knife Ground Fighting

We Hit the Ground
Criminal justice studies report that we hit the ground in a fight 4 different ways. First and foremost we trip and fall. Second we are sucker punched down. Third we are "sport-punched" down. Fouth we are pulled down. These could happen in a mixed weapon fighting. Rule of thumb is get up, unless you can't or you discover a very, unique situational advantage to staying down.

What Kind of Ground?
We fight in rural, suburban and urban area. which is why I am fascinated with the pop term "urban combatives." Some of the best UFC, MMA and Thai fighters are farm boys. But, we fight in these areas and both inside and outside of buildings (and vehicles!). Think of all the types of indoor and outdoor flooring and grounds.

After the Takedown: A Crash Course in Post Event, Weapon Search
Soldier, cop, citizen. You've been in some kind of fight and it's over. You have won, at least so far. But, it's not over until it's over, over, OVER! And over is when the situation is truly over. The suspect must be rendered useless and weaponless. Now you have a suspect or a crazed person, or an enemy combatant on the ground. Restrained, or not? He may still have weapons? His disposition may change and the fight may begin again? He may flee, which would please most citizens but confound professionals. This is not over.

Control, restrain, search and contain. Key words. Key survival skills. Every country has its own official arrest and citizen arrest and search laws. A soldier, enforcement specialist and citizen must familiarize themselves with these laws and procedures as they are unique to each situation and jurisdiction. Aside from the intense and ever-changing search and seizure, legal issues, you first need to ensure your safety and a quick weapon's search is in order. Local search laws do not supercede your immediate safety.

Keep track of his hands. It's his hands that will kill you. It is always easier and safer to search a handcuffed or restrained person. In the field, professionals will have this gear, but the vast population, or undercover soldiers and operatives may not. Use anything in the environment to secure the limbs of the enemy, and behind his back! This could be a supple pants belt, a rope or a drapery string. Anything. Look for these everyday things as a matter of routine.

To start, search the area where you are about to secure the enemy's hands, then secure them. Then begin a search for other weapons. Recall in the Quick Draw section of this book, the three carry sites for weapons.

Primary Quick Draw Carry Sites
These are carry sites that allow for the fastest quick draw. Usually they are around the belt, in and around the pockets of pants and jackets, at times in a shoulder holster type rig. Also, consider the inventiveness of the enemy.

Secondary Quick Draw Carry Sites

These are carry sites where the weapon is somewhat buried under clothing. Usually they are a neck knife rig on the chest, inside shirts or jackets, a boot knife rig. In the armpit inside a bullet-proof vest. In a compartment pocket of a tactical vest. Review the QD photos from that previous chapter.

Tertiary Quick Draw Carry Sites

These are classified as the classic *lunge and reach*, or also *hidden weapons* categories. The enemy might pass a visual or physical inspection, yet may still lunge for a previously concealed weapon in a pre-meditated plan, or a new improvised one. The weapon he lunges for? May also be your own.

Your eyes and often hands must search these three areas. There are two common hand search methods, the pat down and hand crush. Both are self explanatory. With the pat down your fingers and palms feel for weapons on and through the clothing of the enemy. In the hand crush, you squeeze your hands for weapons. In a world of drug addicts and needles and razor blades, enforcement is often warned of these sharp edges as a problem with the hand crush method.

In actuality, searchers probably do both in the course of a common search. Another method used by some professionals is a rub-over search with a small wand about the size of a palm stick. This is not as effective as the use of hands.

And, often criminals have others nearby, such as girlfriends, that hold drugs, guns and knives for them.

Many professionals cordon the suspect's body down into quadrants for searching so they know they will not forget to search a section of a person for weapons, all as a matter of practiced routine.

Once found, collect the weapons and secure them from the enemy, usually on your person. Worried about fingerprints and other CSI evidence? If you are in physical fear, their weapons are less about evidence and more about quick removal for survival. Worry about things like fingerprints later. Leave such weapon removal techniques to calm, seasoned professionals.

Remember you – alive – are the best witness and evidence to this assault case, incident or crime. For the enforcement officer, the soldier? They simply have a new prisoner.

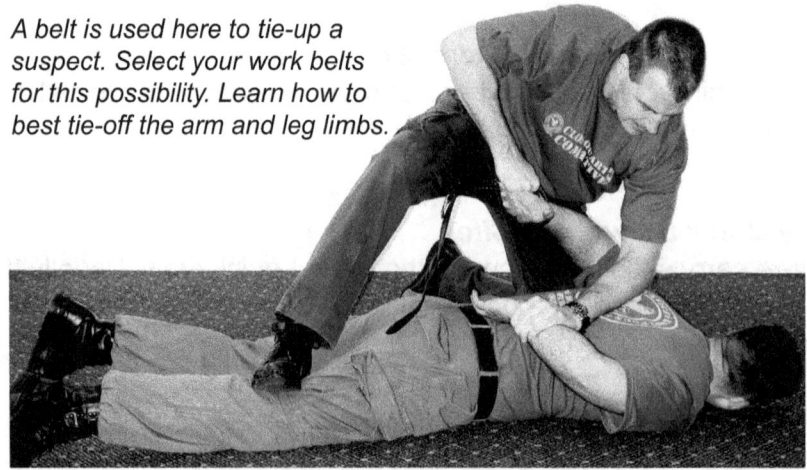

A belt is used here to tie-up a suspect. Select your work belts for this possibility. Learn how to best tie-off the arm and leg limbs.

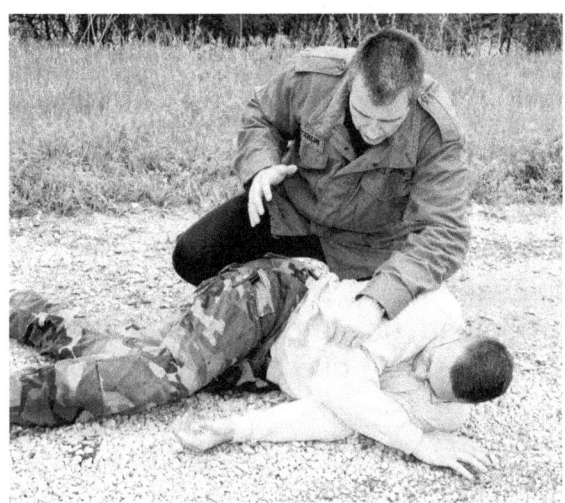
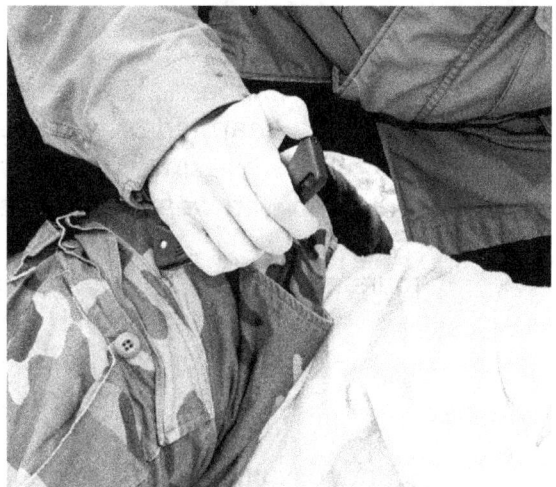

Search the enemy over with your eyes and hands, and especially the primary, secondary and tertiary weapon carry sites for enemy weapons.

"While Holding" a Knife – Ground Maneuvers
For an entire book on footwork and maneuvers, I suggest you get my book "Footwork and Maneuvers for hundreds of how-to photos and explanations. Here we will cover related knife only materials.

Armed Ground Maneuvers

    Grounded Number 1: The Ready Position
        – do unarmed or with weapons sheathed, holstered or mounted.
        – do while drawing a knife, a stick, a pistol, a rifle.
        – do with weapons presented.

    Grounded Number 2: The Shrimp, or Hip Escape
        – do unarmed or with weapons sheathed, holstered or mounted.
        – do while drawing a knife, a stick, a pistol, a rifle.
        – do with weapons presented.

    Grounded Number 3: The Shoulder Walk
        – do unarmed or with weapons sheathed, holstered or mounted.
        – do while drawing a knife, a stick, a pistol, a rifle.
        – do with weapons presented.

    Grounded Number 4: The Bucking Bridge
        – do unarmed or with weapons sheathed, holstered or mounted.
        – do while drawing a knife, a stick, a pistol, a rifle.
        – do with weapons presented.

    Grounded Number 5: The Fishtails (Head and Torso)
        – do unarmed or with weapons sheathed, holstered or mounted.
        – do while drawing a knife, a stick, a pistol, a rifle.
        – do with weapons presented.

    Grounded Number 6: The Sit-Up
        – do unarmed or with weapons sheathed, holstered or mounted
        – do while drawing a knife, a stick, a pistol, a rifle
        – do with weapons presented.

    Grounded Number 7: Guard Rotation
        – do unarmed or with weapons sheathed, holstered or mounted
        – do while drawing a knife, a stick, a pistol, a rifle
        – do with weapons presented.

    Grounded Number 8: The Common Rollover
        – do unarmed or with weapons sheathed, holstered or mounted
        – do while drawing a knife, a stick, a pistol, a rifle
        – do with weapons presented.

Grounded Number 9: The Scissor Kick Rollover
- do unarmed or with weapons sheathed, holstered or mounted
- do while drawing a knife, a stick, a pistol, a rifle
- do with weapons presented

Grounded Number 10: Side Rotation, The Curly Shuffle
- do unarmed or with weapons sheathed, holstered or mounted
- do while drawing a knife, a stick, a pistol, a rifle
- do with weapons presented.

Grounded Number 11: Back Pivot Rotation
- do unarmed or with weapons sheathed, holstered or mounted
- do while drawing a knife, a stick, a pistol, a rifle
- do with weapons presented.

Grounded Number 12: Hip Pivot Rotation
- do unarmed or with weapons sheathed, holstered or mounted
- do while drawing a knife, a stick, a pistol, a rifle
- do with weapons presented.

Grounded Number 13: Get up!
- from on back.

Rattlesnake Overview

Material involving ground fighting could fill volumes. Add knives into the fight and combat becomes complicated and more deadly. The purpose of this section is to establish some basic positions, responses and variables in which a soldier, Marine, security, law enforcement and civilians might find themselves involved.

When people hit the ground in a knife fight, it is perfectly common for them to recognize the obvious dangers and instinctively move apart. This is usually done by kicking and shoving the opponent with the legs. But, not so, with some martial artists trained in sport fighting. Over the years I have observed many knife killshot tournaments and quite a number of participants accidentally fall to the floor. I have seen these brainwashed sport-wrestling training habits put the practitioners in suicidal ground fighting positions! Time and time again, martial art ground fighters competing in hard core knife tournaments who tackle, or trip and fall in battle have this compulsion to wrap each other up closely and proceed to stab each other to simulated death. With our Killshot rules, if there is no winner, then both participants die and nobody wins. Non-martial artists and non-ground fighters avoid this ground clinch and snake away from each other's blades.

There has been a martial obsession for years now that has poisoned even military combatives into sport, ground fight wrestling from the mandated, classic sport guard and mount positions, the lack of no rules, cheating.

There has been a martial arts obsession for decades now that has poisoned survival ground fighting even seeping into military combatives where combatants thoughtlessly resort to sport, ground fight methods and positions from the mandated, classic sport guard and mount positions.

There has been an obsession to promote the classic sport positions of the "guard" and the "mount" as some cure-all for all ground zero problems. Students mindlessly roll around and practice these ideas for hours and hours, year after year until sport wrestling methods become second-nature under all circumstance. Remember, when under serious stress, we drop to our repetition training. Lost are:

– all cheating in sport wrestling:
  * hair pulls.
  * face mauls.
  * bites.
  * eye attacks.
  * in most cases, finger and wrist attacks.
– strikes to common and illegal places.
– kicks.
– mixed weapons.

Sport Wrestler's Mistake. The knife-fighting suicide pact. "Let's just tie-up close and kill each other."

From organizing and refereeing these bouts, it became apparent to me years ago that hitting the ground and wrapping your arms and/or legs around each other in a knife fight is little more than a suicide pact. But you may end up this way!

Sport ground fighters, even those within the current and popular trends of Mixed Martial Arts, and who think they are doing absolute reality fighting, have developed this muscle memory to hug, pin and control their opponents, especially with their legs, not to use their legs to kick someone away, or be free in thought and action to improvise. There may well be a time and a place for the full wrestling applications. But knife ground-fighting isn't the time.

Improvise! You need to keep your options open for review and selection. What if you are the only one armed with a knife, and he has none? If you have been attacked and in fear for your life, you may see fit, due to your confidence, skill and the mission-at-hand, to leg tie your unarmed attacker in close to you. Now he is captured and not easily able to elude your blade. If you are unarmed against the knifer, you should try to evade as you hit the ground. Your muscle-memory cannot let you tie your legs around him and keep him near.

When you are the only one armed with a knife, you may see fit, due to your confidence, skill and mission, to leg-tie your unarmed enemy in very close to you. Now he is captured and not able to easily elude your blade.

Sport Wrestler's Mistake. Here, the bottom-guy is unarmed. Top guy is armed. The bottom-guy mindlessly resorts to his sport muscle memory and leg ties the knifer to him. He pulls himself in and close to the enemy's knife.

When people hit the ground in a knife fight, it is perfectly common for them to recognize the obvious dangers and instinctively move apart. This is usually done by kicking and shoving the opponent away with the legs. We call this a "raised shin guard."

In summary, combatives practitioners study all types of ground fighting, even sport, to learn to defeat it, not become it. Practice the smarter, survival scenarios over and over again. Remaining free from sport routines means remaining free to improvise. Improvisation is a key skill in overcoming the chaos in a knife ground fights and all fights.

Ground Tactic Extraction Study – Being Caught in Armbars

Extract the finer points and applications of sport wrestling that may work in knife ground fighting. An innovative, enlightened, and veteran trainer can assess what will and will not work. Below is a chart to emphasize the point. The big box represents all of submission ground fighting. Much of this material would not be effective against a rabid knife attacker. Yet, the aspects that would potentially work, are the arm bars and arm smothering skills of a good submission system. These, with reality add-ons, could be made to work in a knife ground fight. To say that none of it will, ignoring and avoiding all of it, is a tactical mistake.

It is becoming more and more evident with the success of Mixed Martial Arts *ground n'-pound*, that is heavy ground striking and kicking, that submission wrestling is not a stand-alone, reality fighting system, but rather a sport with rules.

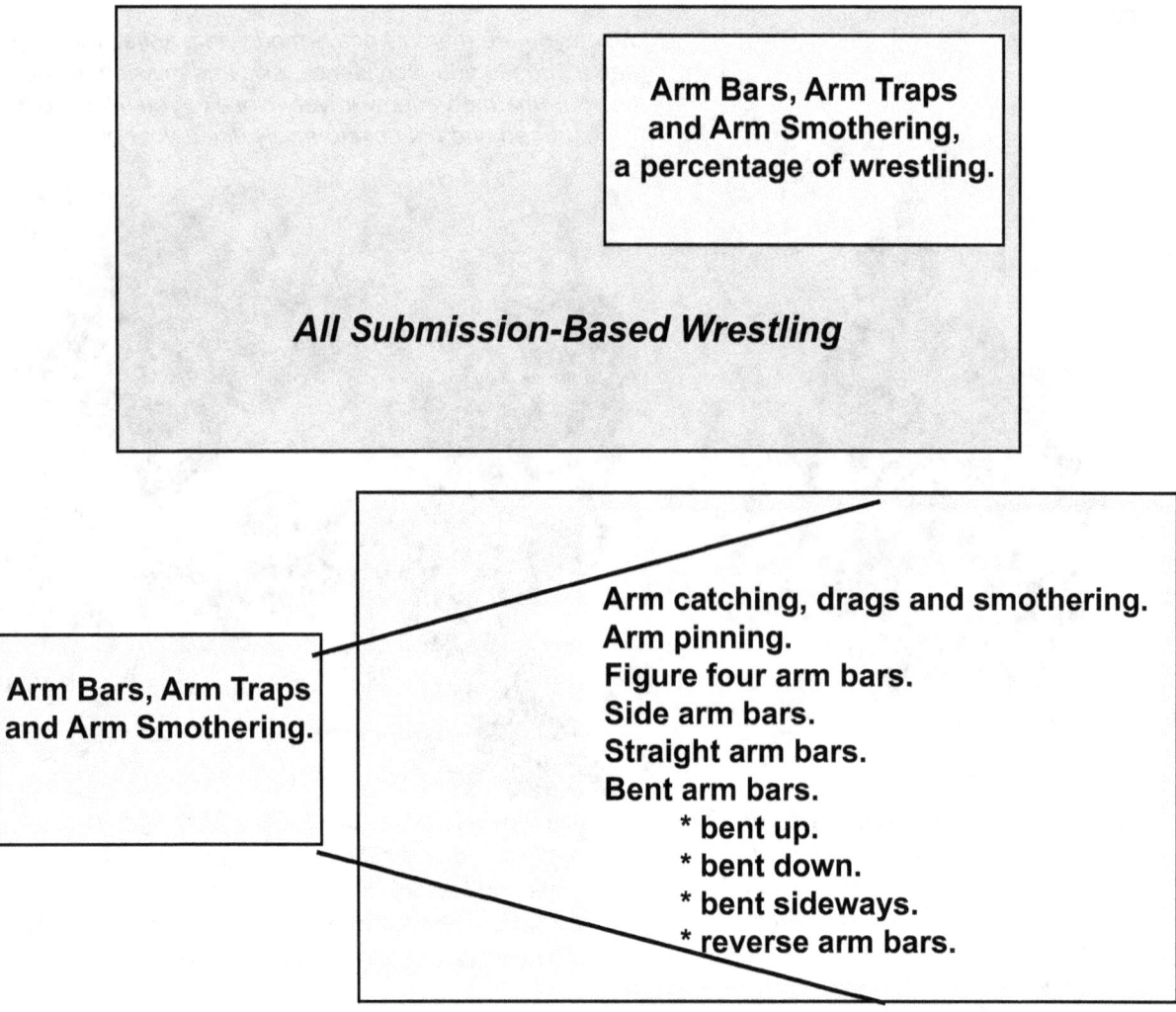

Rattlesnake Basic Positions

First, basic knee positions are established:
- Ground Position 1: Knee high fighting standing.
- Ground Position 2: Knee high fighting knee high.
- Ground Position 3: Knee high fighting grounded.
- Ground Position 4: Topside.
- Ground Position 5: Bottom-side.
- Ground Position 6: Side-by-Side.

The Knee-High Position Samples

 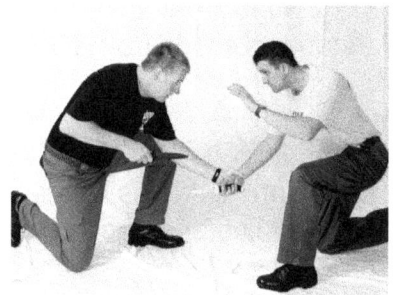 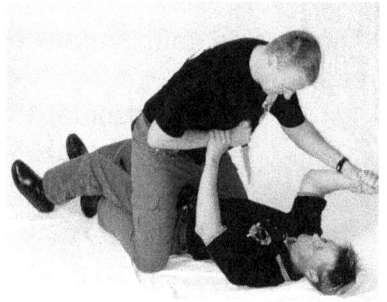

Knee-high vs. standing.   Knee-high vs. knee-high.   Knee high vs. grounded.

* The Topside Knee Positions

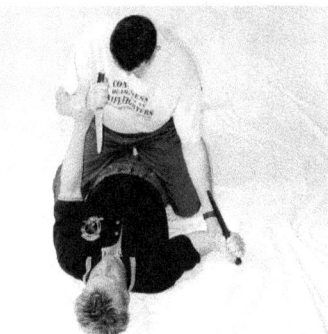 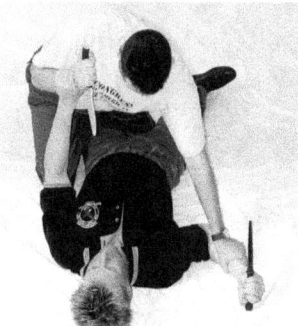 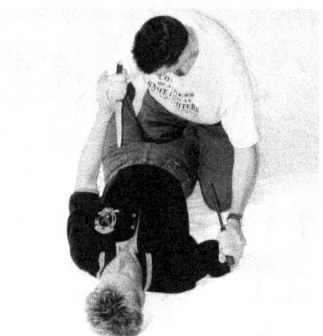

Topside and both knees out.   Topside and right knee out.   Topside and left knee out.

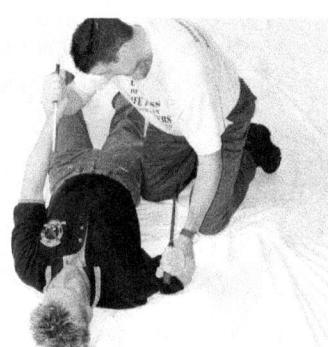

Topside Knee Possibilities
1: Both knees spread outside the body.
2: Left knee inside enemy's legs.
3: Right knee inside enemy's legs.
4: Both knees inside enemy's legs.
5: Both knees outside enemy's legs.
6: Right knee up, left down.
7: left knee up, right down.
8: Both knees down.

Both knees outside. Right or left side.   Both knees inside the opponent's legs.

Vertical and Horizontal Range "Scales"

The topside defender must understand methods from each of the three, knee-high vertical positions:

Vertical 1: High.
Vertical 2: Medium.
Vertical 3: Low.

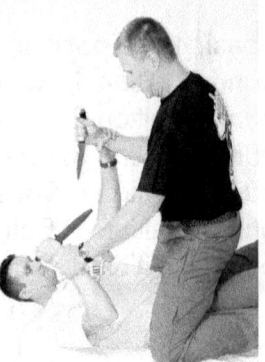 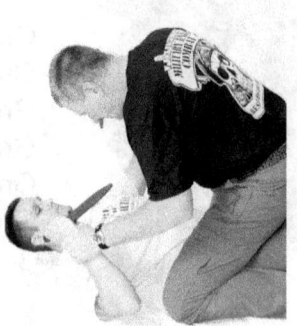 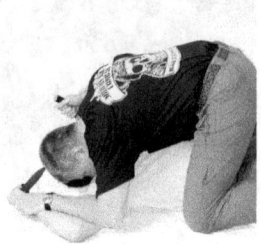

High vertical.   *Medium vertical.*   *Low vertical.*

The topside defender must understand methods from each of the three horizontal knee positions.

        Horizontal 1: Knees high.
        Horizontal 2: Knees medium.
        Horizontal 3: Knees low on hips. This acts as a pinching guard against weapon belt line and pocket weapon quick draws.

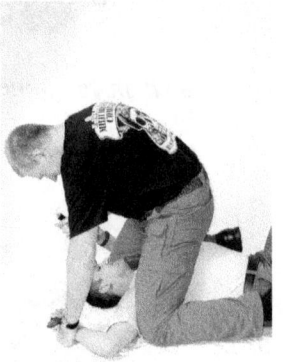 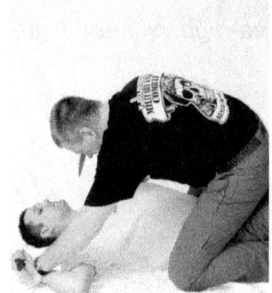 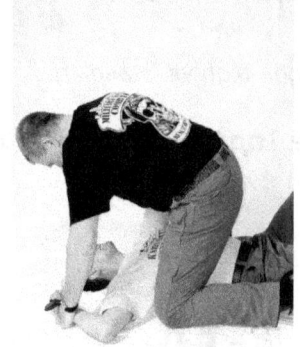

Knees high on shoulders.   *Knees medium on the torso.*   *Knees low on the hips.*

Review all topside, knee-high knife draws, strikes, blocks and passes.

1: All thrusting knife attacks.
2: All hooking knife attacks.
3: All saber and ice pick slashes.
4: All support strikes and kicks.

* Other Sample Topside Knee Situations

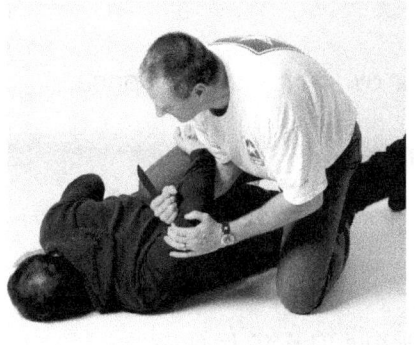
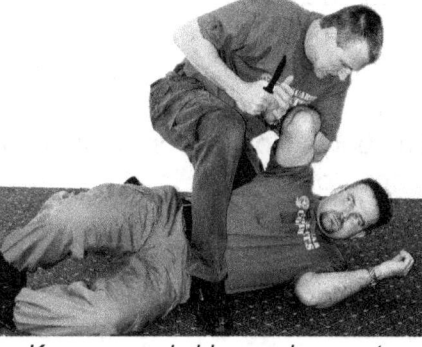
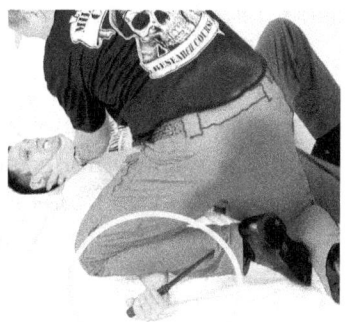

Knee on back and arm bar capture.   Knee up and side arm bar capture.   Knee and shin trap.

* Bottom-side Ground Situations

Review the top-side photos and examine the bottom-side person for more examples of being on the bottom-side of a ground fight.

* Side-by-Side Ground Sample Situations

At times, this raised right leg can reach over the soldier's weapon-bearing limb and help pump the weapon downward and into the enemy's weapon limb.

The Rattlesnake Quick Draw Exercises
Get "knocked down" or tackled each time.

* Grounded draw practice, no trainer. Trainees draw at their own experimental pace.
   – trainee grounded knee high.
   – trainee grounded on back. No cues or training partner.
   – trainee grounded face down.
   – trainee grounded on right side.
   – experiment with ankle sheath draw.

* Grounded with standing trainer. Trainer cues trainee with pulling a knife or other deadly weapon.

* Grounded on back, with trainer on top. Top-side trainer with boxing gloves offers stress to trainee's knife quick draw.

* The Ground Shield scenario. Trainee is struck down. Trainer draws knife and approaches. Trainer shoulder walks for space, maybe bicycle kicks for space, draws knife, slashes in the air moving trainer back. Trainee gets up.

* Rattlesnake Quick-Stress Knife Draw Drills:
These three sets constitute a great deal of knife ground fighting by themselves.
   Set 1: Trainer ground fights, "unaware" of knife draw and offers trainee simple stress.
      * trainee top-side and draws.
      * trainee bottom-side and draws.

   Set 2: Trainer ground fights, tires to stop knife draw.
      * trainee top-side and draws.
      * trainee bottom-side and draws.

   Set 3: Trainer and trainee both draw and both try to stop each other
      * you topside.
      * you bottom-side.

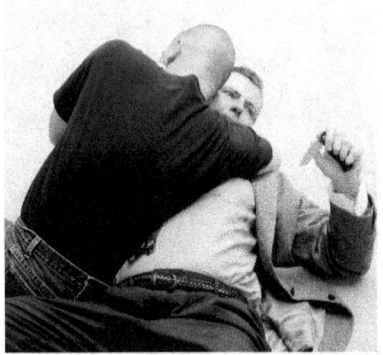 

It is always the trainer's job to develop the trainee.

Soldiers, enforcement and citizens must practice stress quick draws from all these ground positions. If your mission is dangerous, it makes sense to carry two knives, one on each side. When ground fighting, one never knows what side will be pinned into the ground, sealing off weapon access.

Knife Ground Fight Basics: Kicks

Rattlesnake Kicking
The defender can kick the enemy or use his feet to push the enemy's body away. The defender can use the his legs for kicking, kneeing or pinning.
- thrust kicks or knees.
- hook kicks or knees.
- snap kicks or knees.
    * the lower legs.
    * the thighs.
    * the groin.
    * the stomach.
    * the chest.
    * the face.
    * the far arm.
    * traps.

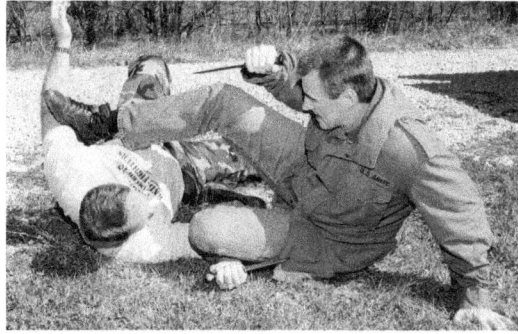

Knife Ground Fight Basics: Stabs, Slashes, Strikes
Review all prior knife attacks, blocks and support tactics.

Knee High
Saber grip thrusting stabs.
Saber grip hooking stabs.
Saber grip slashes.
Reverse grip thrusting stabs.
Reverse grip hooking stabs.
Reverse grip slashes.
All blocks and hacks.
All pommel strikes.
All support hand strikes.
All customized kicks for this position.

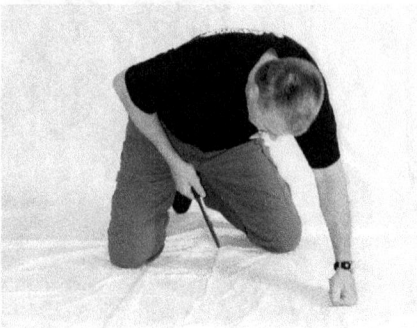

Topside
Saber grip thrusting stabs.
Saber grip hooking stabs.
Saber grip slashes.
Reverse grip thrusting stabs.
Reverse grip hooking stabs.
Reverse grip slashes.
All blocks and hacks.
All pommel strikes.
All support hand strikes.
All customized kicks for this position.

Bottom-side
Saber grip thrusting stabs.
Saber grip hooking stabs.
Saber grip slashes.
Reverse grip thrusting stabs.
Reverse grip hooking stabs.
Reverse grip slashes.
All blocks and hacks.
All pommel strikes.
All support hand strikes.
All customized kicks for this position.

Side-By-Side
Saber grip thrusting stabs.
Saber grip hooking stabs.
Saber grip slashes.
Reverse grip thrusting stabs.
Reverse grip hooking stabs.
Reverse grip slashes.
All blocks and hacks.
All pommel strikes.
All support hand strikes.
All customized kicks for this position.

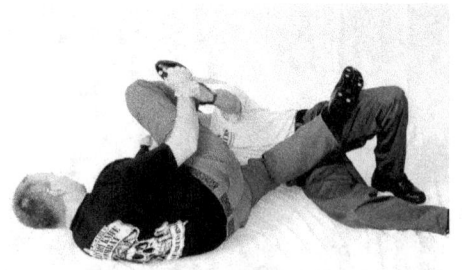

Knife Ground Fighting Combat Clock

Possible Positions Overview

Topside or bottom-side in a ground fight, knees up, knees down or flat, you will find yourself in the Combat Clock positions. It is a good idea for the defender to familiarize with each one and make a plan:

- Right knee up / left knee down.
- Left knee up / right knee down.
- Both knees down.
- Both knees on the right side.
- Both knees on the left side.
- Your knees inside his legs.
- His legs inside your knees.
- Your right knee between his legs.
- Your left knee between his legs.

Basic Training

Many ground fight systems satisfy themselves with the four corners approach using the terms *North, South East, West*. But this leaves no detailed map for more specific positioning study. The Combat Clock does allow for this growth. Starting with the horizontal, Combat Clock, Basic Training begins with and understanding of the four basic positions:

- 12 o'clock.
- 3 o'clock.
- 6 o'clock.
- 9 o'clock.

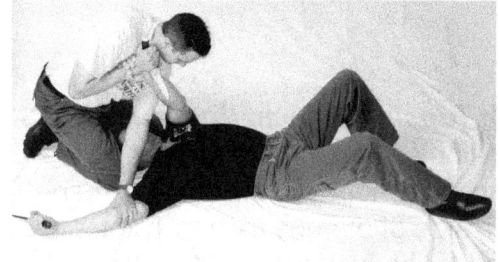
Basic 12 position or high.

Basic 3 o'clock position or right side.

Advanced Training

Much training can be accomplished with the four basic corners, but when needed the advanced training provided by using all the numbers of the combat clock allows doctrine to be thorough and complete.

Familiarize the defender with all 12 numbers of the Combat Clock in ground positions.

- all 12 numbers of the Combat Clock.

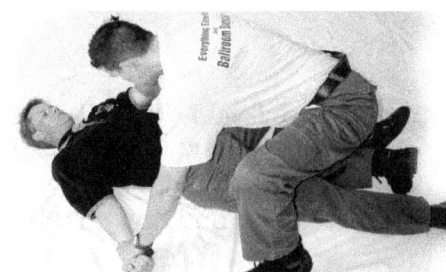
Basic 6 position or low.

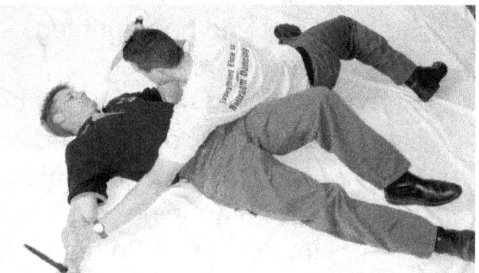
Basic 3 o'clock position or right side.

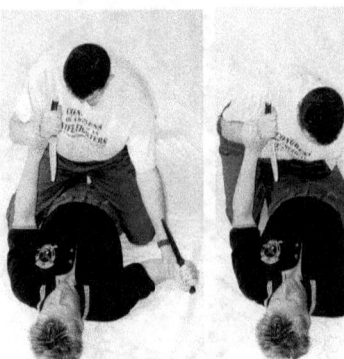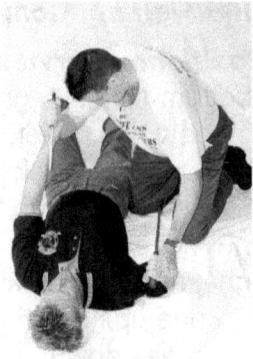

Both knees out.   Left knee in.   Right Knee in.   Both knees out and to one side.

The Topside Ground Post and Hop Skill Drill

Position change in ground fighting is a proven, valuable skill. This Post and Hop Drill simply develops skills in topside maneuvering. The trainee starts with two knees to the right side of a prone trainer. The trainer holds a knife. The trainee posts a hand, or a knife hand, or both on the chest and/or face of the trainer. The Trainee proceeds to leg hop across the legs of the trainer. The trainer may start with his legs flat, but slowly raise them to create a needed, realistic "hop" to clear them. Imagine the damage done by a posted, saber or ice pick grip knife into the enemy's body.

Postings
    Post 1: Free hand.
    Post 2: Knife hand or knife into body.
    Post 3: Both hands.
    Pivot: Pivot chest-to-chest. The leg still "hops" to get into position.

The trainee leg hops though each of these positions. His hands, forearms or elbows can post on the body and/or the ground. Or, the trainee can practice pivoting on the trainer's chest.

Drill Steps
    Step 1: Starting Position.
    Step 2: Post and hop left leg over trainer's left leg.
    Step 3: Post and hop right leg between trainer's legs.
    Step 4: Post and hop left leg over trainer's right leg.
    Step 5: Post and hop right leg over trainer's right leg.
    Step 6: Post and hop right leg over trainer's torso.
    Step 7: Post and hop right leg over trainer's right leg.
    Step 8: Back to starting position.
    Step 9: Try no hopping. Lay chest to chest and spin legs over the trainer's head to get both knees on the starting side.

The Topside Ryan Combat Scenario Series

Since well before the movie *Saving Private Ryan* was made, I have been teaching topside knife fighting tactics. Since the 1990's movie, these topside, ground moves have been nicknamed the *"Private Ryan Series."*

Ryan 1: Stabbing Ryan – Use body weight to drop the knife into a chest stab.
Ryan 2: Climbing Ryan – Use legs to pin arms.
Ryan 3: Releasing Ryan – Use all the releasing tactics to get hands free.
Ryan 4: Dismounting Ryan.

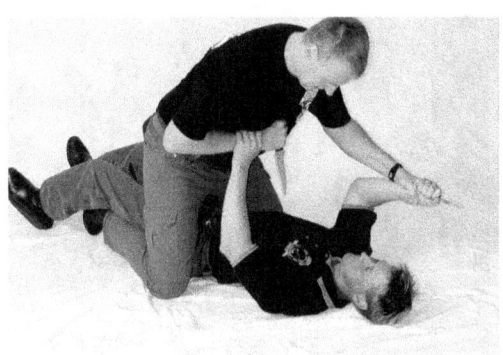

1: Stabbing Ryan:
No matter your angle, steer the knife handle to your chest, aim at the enemy's chest, neck or face of the enemy and press down as post/base for a leg hop or body spin.

2: Climbing Ryan:
The defender immediately knee and shin traps the trainer's left arm. This is painful. Then, he pulls free from the enemy's grip and attacks the enemy with his newly freed knife.

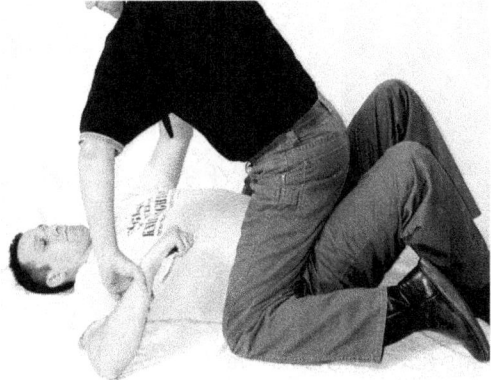

Ryan 3: Releasing Ryan
Use all the releasing tactics to get hands free.
– stab the arms.
– slash the arms.
– circular releases.
– yank-outs.
– bite hands, arms, face, and neck.

Ryan 5: Dismounting Ryan (formally the Paladin Dismount)
If a defender suddenly sees the need to escape from a topside position, he can, and still render a surprise, fatal strike. The defender maintains a tight grip on the enemy's wrist. He executes a forward roll, completely escaping the position and confusing the enemy with an unusual tactic. As soon as he hits the ground, he spins and stabs violently at the enemy's face and throat, from a surprise angle.

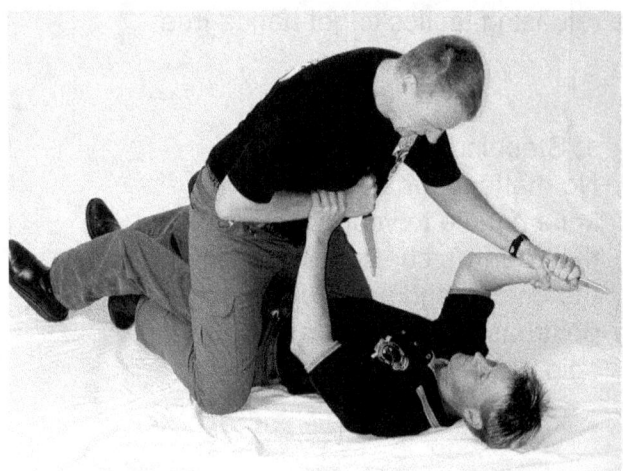

This begins at classic topside position. The defender points his knife at the heart area of the enemy and...

The defender hits the ground, spins and attacks the neck. The defender never let go of the enemy's weapon limb. (Photos taken while teaching US Marines combatives at Quantico, VA.)

Rattlesnake Knife Combat Scenario Sample: The Quick Draw exercises
Probably one of the most important ground knife exercises you can do.

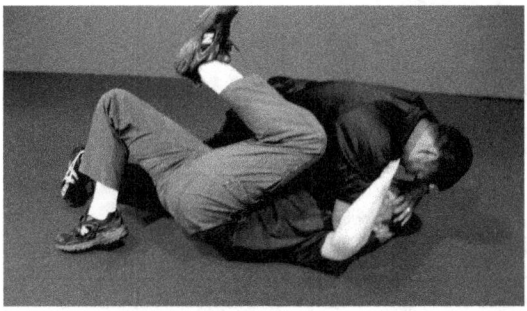

First set – trainer just offers general stress.

Second set – trainer offers general stress as well as tries to interfere with trainnee knife draw.

Third set – both try to draw knives and stop the other from drawing.

Rattlesnake Knife Combat Scenario Sample: The Shield
A sample of another stress knife, ground quick draw scenario.

 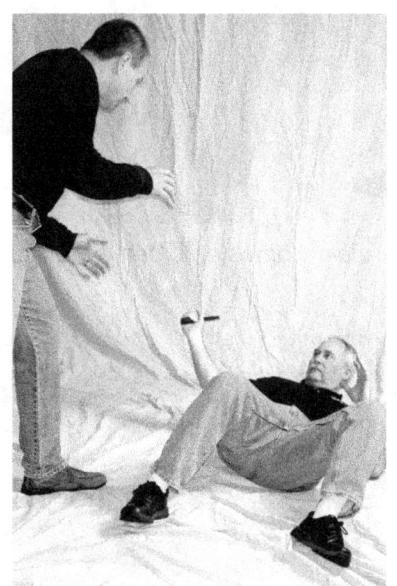 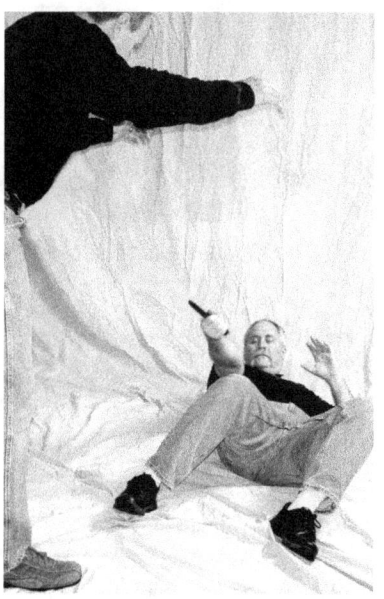

The defender slashes away to attain the space to get up.

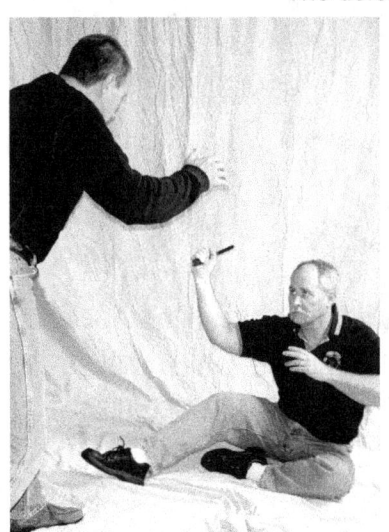  

The defender slashes away, creating a shield, to attain the space to get up.

Rattlesnake Knife Combat Scenario Sample: Slipped Down But Not Out

The civilian, disguised soldier or plain clothes enforcement officer slips back while in a lethal fight.

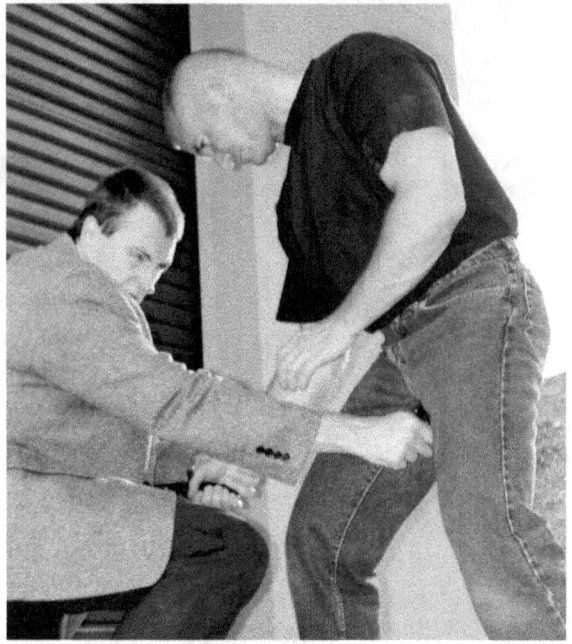

Low targets! Thighs. Groin, Stomach. Here, first the groin is stabbed, which will usually cause the enemy to buckle and bend, bringing the stomach into target range. Multiple, rabid knife attacks should cause the enemy to back off.

Rattlesnake Knife Combat Scenario: The Big Stab Down

At times the assailed must fight from grounded positions because they have been wounded, or have fallen. In the sample, an enemy stabs a victim in the stomach as the victim was drawing his knife in self defense. Few knife systems cover post-wounding responses. Think about fighting back from one.

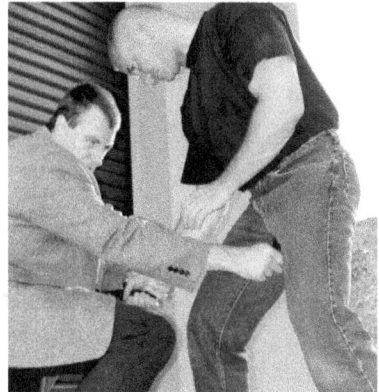

Stabbed! Painfully and sudden. The victim drops to one knee. The enemy withdrawals his blade. The victim follows the withdrawal with his own knife and then hook stabs deep into the groin with a powerful uppercut.

The enemy doubles over from the groin stab. The victim's knife edge rakes the forearm as he pulls the enemy down and over.

The wounded victim moves in to finish the conflict as needed. Fight through the pain and shock to survive. Next, self treatment and medical attention.

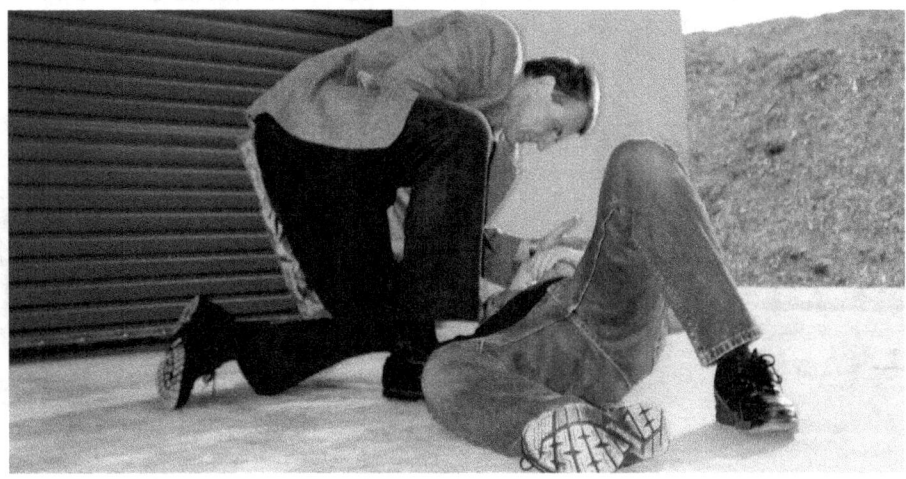

Rattlesnake Combat Scenario: Pinch-off the Quick Draw

The defender may fight an enemy wearing web or gear belts. Their belts hold pistols and knives.

In such a combat situation, the defender is knocked down, or has fallen with the enemy. The enemy is topside and between the defender's legs.

The enemy decides to draw his sidearm. This exact motion could also draw a knife from a sheath on a gear belt, or from a pocket.

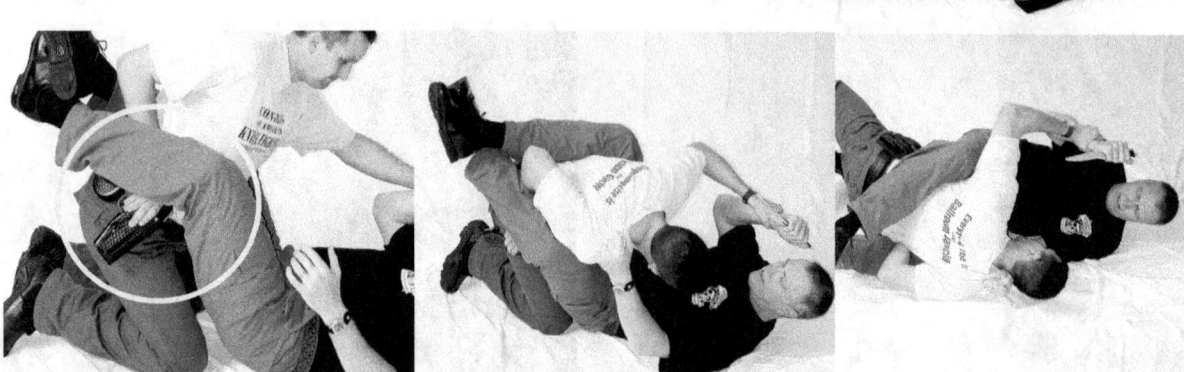

The defender hooks ankles and knee pinches the weapon grab. Extending the ankles increases the power of the pinch, catching the bulky pistol and grab. The defender may have to adjust the leg position for maximum catch pressure. The defender may strike the face. The defender "power-shifts" his hips to his right, grabs the enemy's torso and drives the enemy over and down atop his weapon side. The defender tries to slip his bottom-side leg out, while maintaining the most powerful pinch on the weapon.

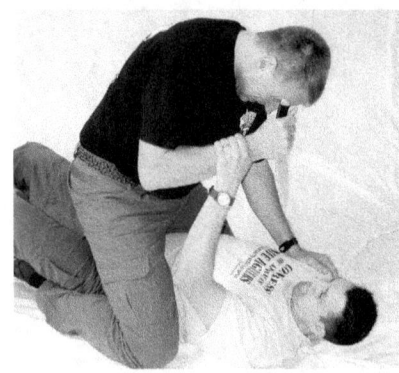

It is possible to still contain the weapon, the weapon hand and weapon wrist within the knee pinch. If so, then slap the enemy's grip on the defender's knife-side limb. Attack as needed in the situation.

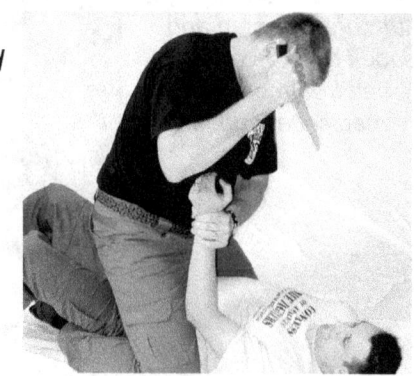

Rattlesnake Knife Ground Combat: The Basic Body Bump

Hip movement is a critical aspect of ground fighting. And one of the most simple, yet effective methods for getting a topside person off of you in a ground fight is with a powerful hip movement. This movement is called several different things in martial arts, a "Pelvic Bump," a "Pelvis Bump," even a "Hip Heist," combined with a power hip shift to the side that is called a "Shrimp" by many experts because the soldier's body will end up in a curved position. Each one of these terms might have slightly different meanings in different systems. This movement, along with upper thigh and knee strikes may also help move opponents around. Every defender should be familiar with this high yield, bottom-side, basic, ground movement.

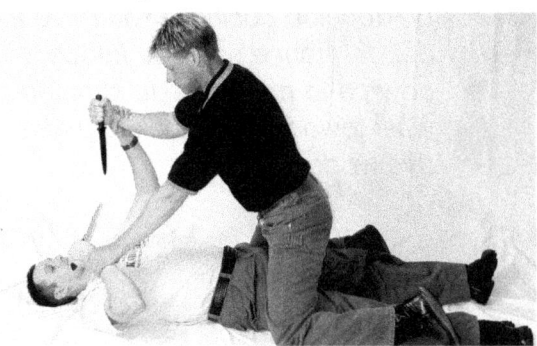

The defender is downed and the enemy has jumped on him.

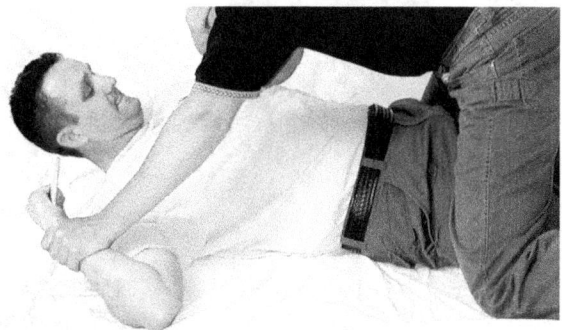

The defender begins a circular releasing move to free his knife limb.

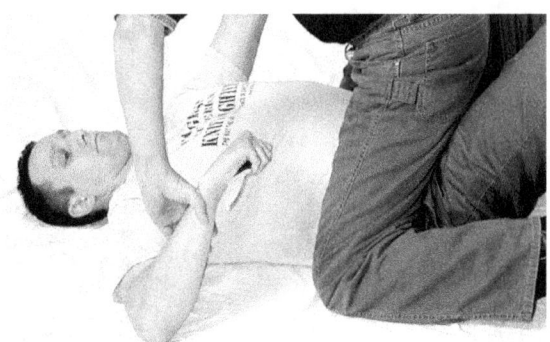

The defender continues the circle. he holds the enemy's right arm. He plants his feet.

The defender could stab and push the torso. Here he uses the side of the blade to push. Note the planted feet and hip rise. Hip emphasis to the high left.

He pitches the enemy over, while still holding the enemy's weapon limb.

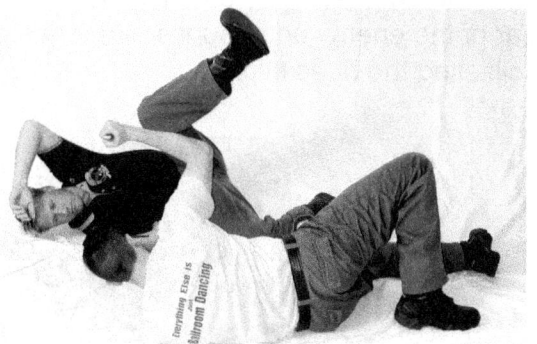

He shifts his hip to the right. Once the enemy is flipped, the defender attacks the best targets he can see. The wrestling is over!

Rattlesnake Knife Combat Scenario: The Splitter

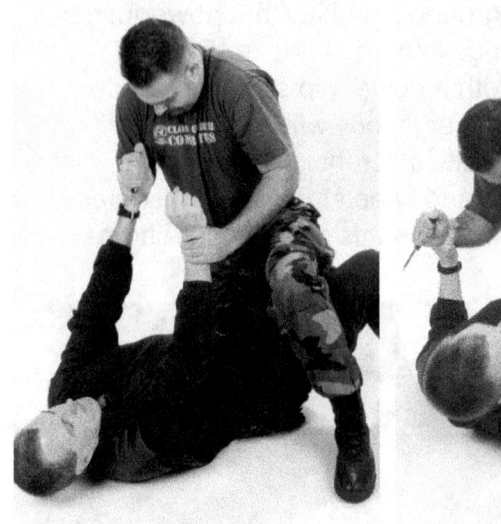 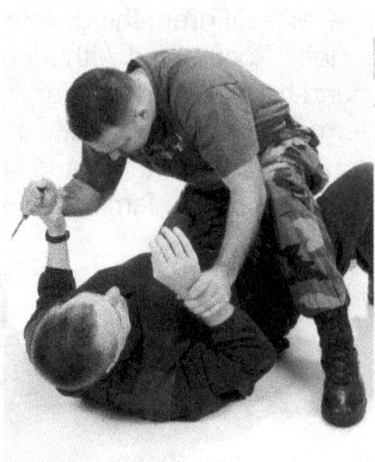

The defender is on the bottom-side of a topside knife attacker. The enemy is overpowering and forces the defender's arms down and spread apart – thus the name – *The Splitter*.

(Note the common "one-knee-up" topside position. You have a better chance with two knees down and may have to shoulder-walk toward 12 o'clock to make this work.)

The arms are split. The defender bites the enemy's arm if possible (think clothing?). The defender slides this captured arm up and over. He slips his head under the arm. The defender may use his knee/thigh to strike the rear of the enemy and bounce him forward too, facilitating the head slip.

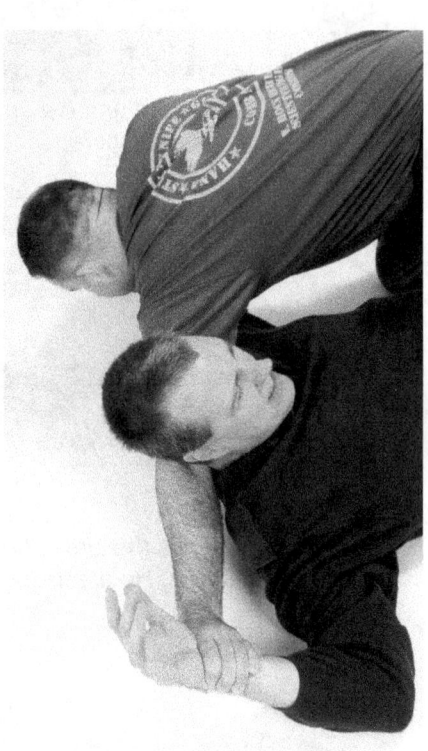

With this "arm-bar-bridge," the head pressures the enemy elbow. The soldier's body arches and he yanks his captured hand free.

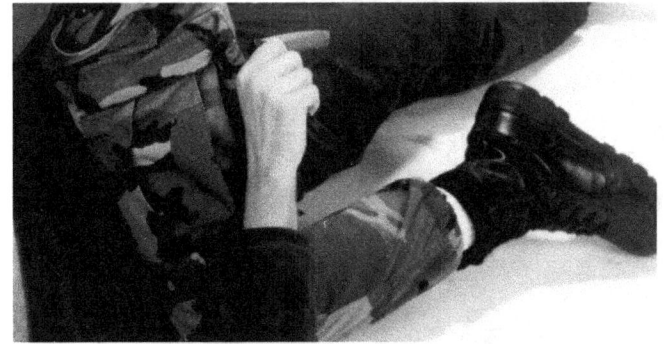

The defender accesses his knife from his belt line or pocket. In this photo series, he must open a tactical folder. This access may be from the space between his right leg and the soldier's left leg, or the soldier may have to reach around the outside of the enemy's leg to his knife.

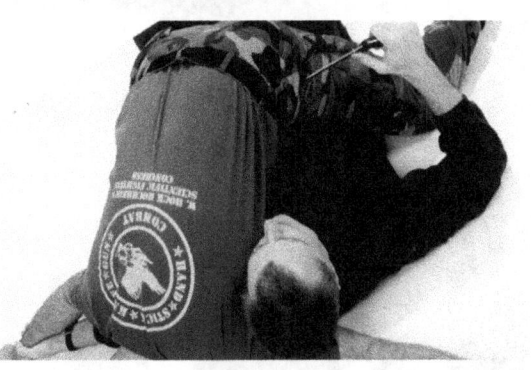

Meanwhile the defender tries to push the enemy's arm against his head. The soldier stabs into the enemy's torso. Multiple times. Perhaps even announcing the event because he cannot see the attacks.

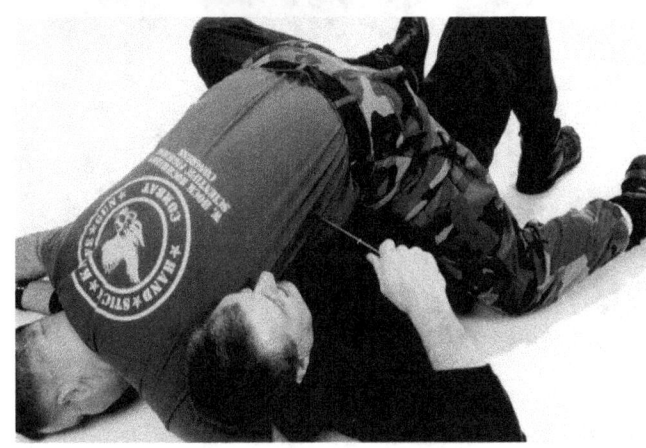

The defender plants his foot flat for support, arches his back, lifts his pelvis and twists his body to throw the enemy off of him, all the while pushing his knife into the torso. Next, a side-by-side combat scenario continues.

Push off and kick like mad. The wrestling is over. Do not roll over atop the attacker fore more and more wrestling.

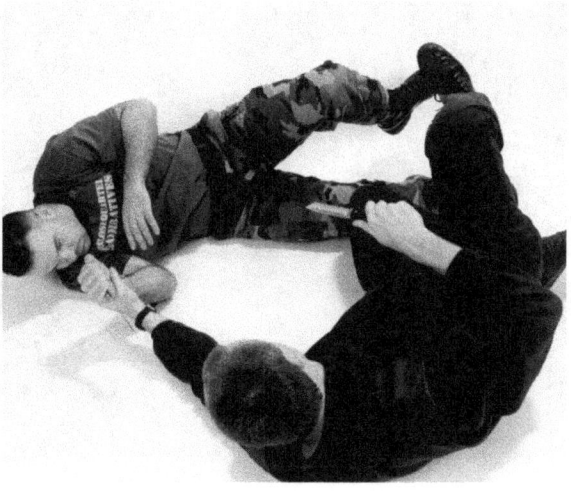

Rattlesnake Knife Combat Scenario: A Circular Release Sample.

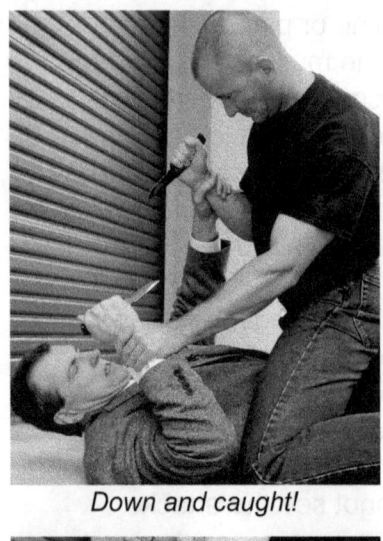
Down and caught!

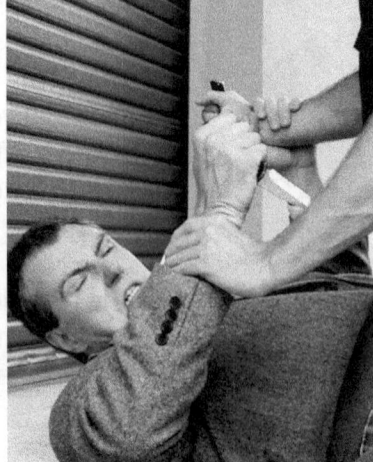
Wound his defense.

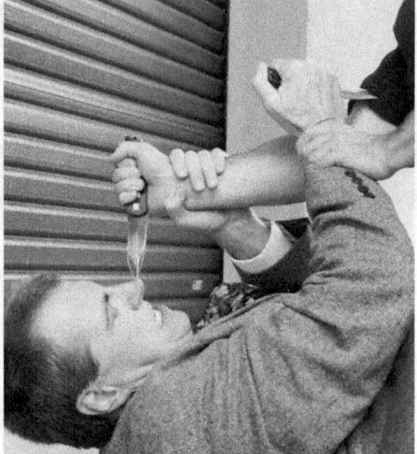
Wound his offense.

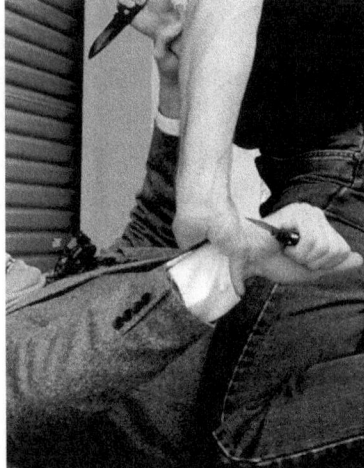

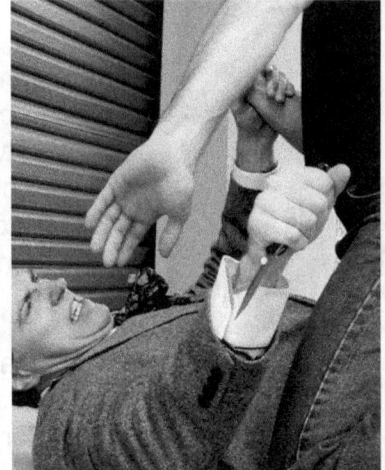

A circular release, then attack the enemy's body. Further tactics for this situation will follow.

Rattlesnake Knife Combat Scenario: A Joint Lock Release Sample.

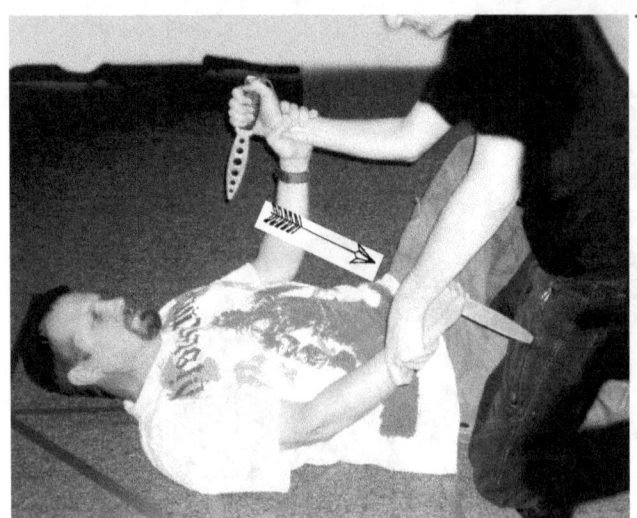

The defender is down and caught, his knife hand grabbed. Wound his defense and wound his offense as shown above. The defender brings his hand and knife in a bit to create an awkward grip, then he stabs down into the attacker's thigh.
Circle out for more moves.

Rattlesnake Side-by-Side Combat Scenarios

Side-by-side (SBS) ground fighting may truly be a unique event and position in all ground fighting. Unarmed fighters usually move to tighter positions. But when the two fighters have knives and hit the ground landing sideways, they tend not to let go of each others weapon limbs, and prefer to keep their distance. They often shove each other with their feet to gain some physical SBS distance.

Though grounded, the combatants are essentially in *"The Death Grip Of"* each other, and may use many of the problem-solving methods from that prior chapter in this book. Unique elements to this ground fight:

SBS 1: The defender may have the top-side knife, as the defender on the left demonstrates.

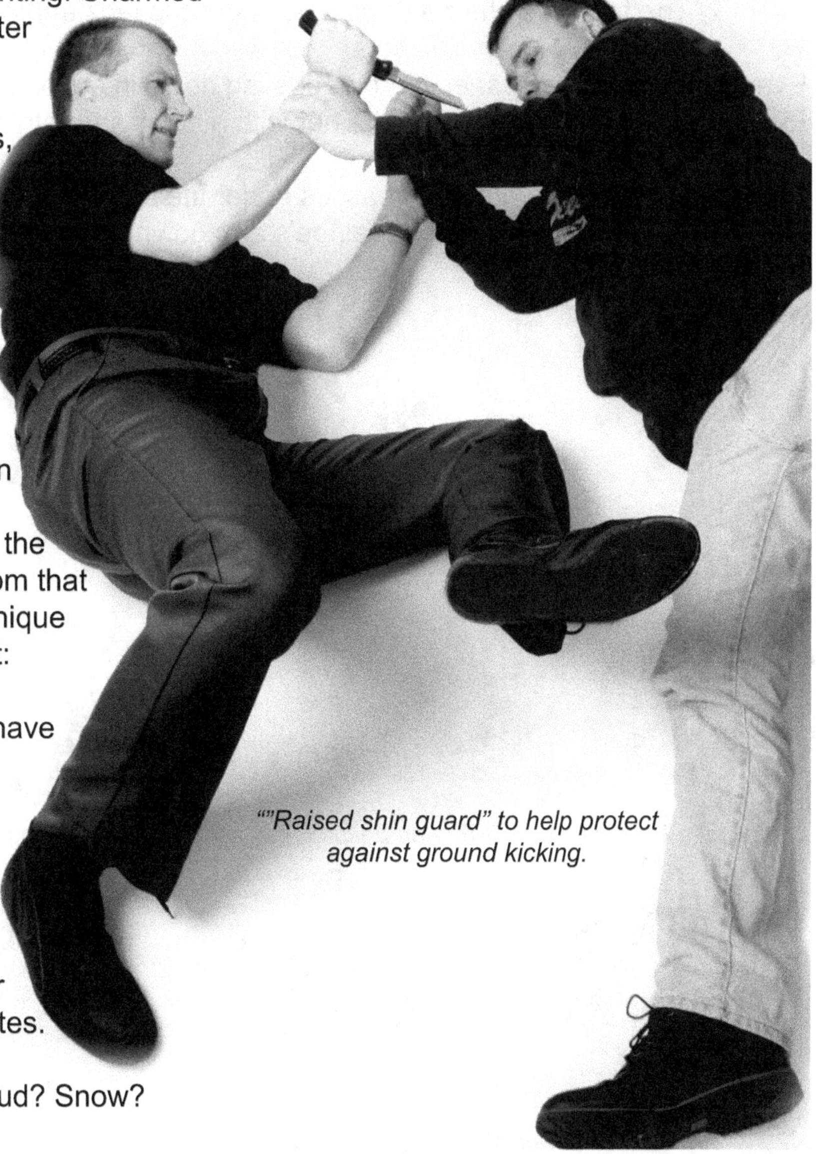

""Raised shin guard" to help protect against ground kicking.

SBS 2: The defender may also have his knife on the bottom-side, as the attacker on the right side demonstrates.

SBS 3: Weather. Water? Mud? Snow?

SBS 4: Ground textures and conditions. Where are the fighters fighting? Atop rocks? Cement? Debris? Interior floors?

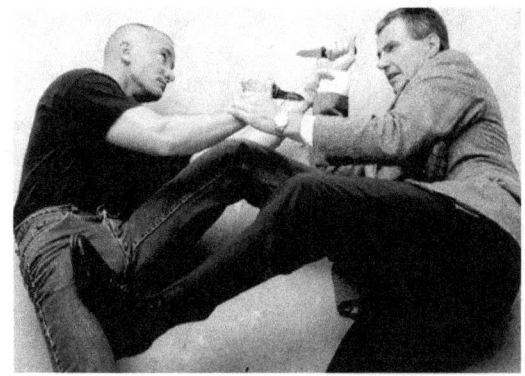

The Rattlesnake Block, Pass and Pin Exercise

We repeat this here so that this chapter is stand-alone thorough.

Event 1: The trainer kicks downward at an angle. Trainee blocks.
Event 2: The trainee's other arm passes under the legs and pushes off the trainer's leg.
Event 3: The trainee pushes the trainer's leg down in a pinning fashion.
Event 4: The trainee attacks the trainer with a downward strike at an angle. Trainer blocks.
Event 5: The trainer's other leg passes under the leg and pushes off the trainee's arm.
Event 6: The trainer pushes the trainee's arm down in a pinning fashion.
Repeat...

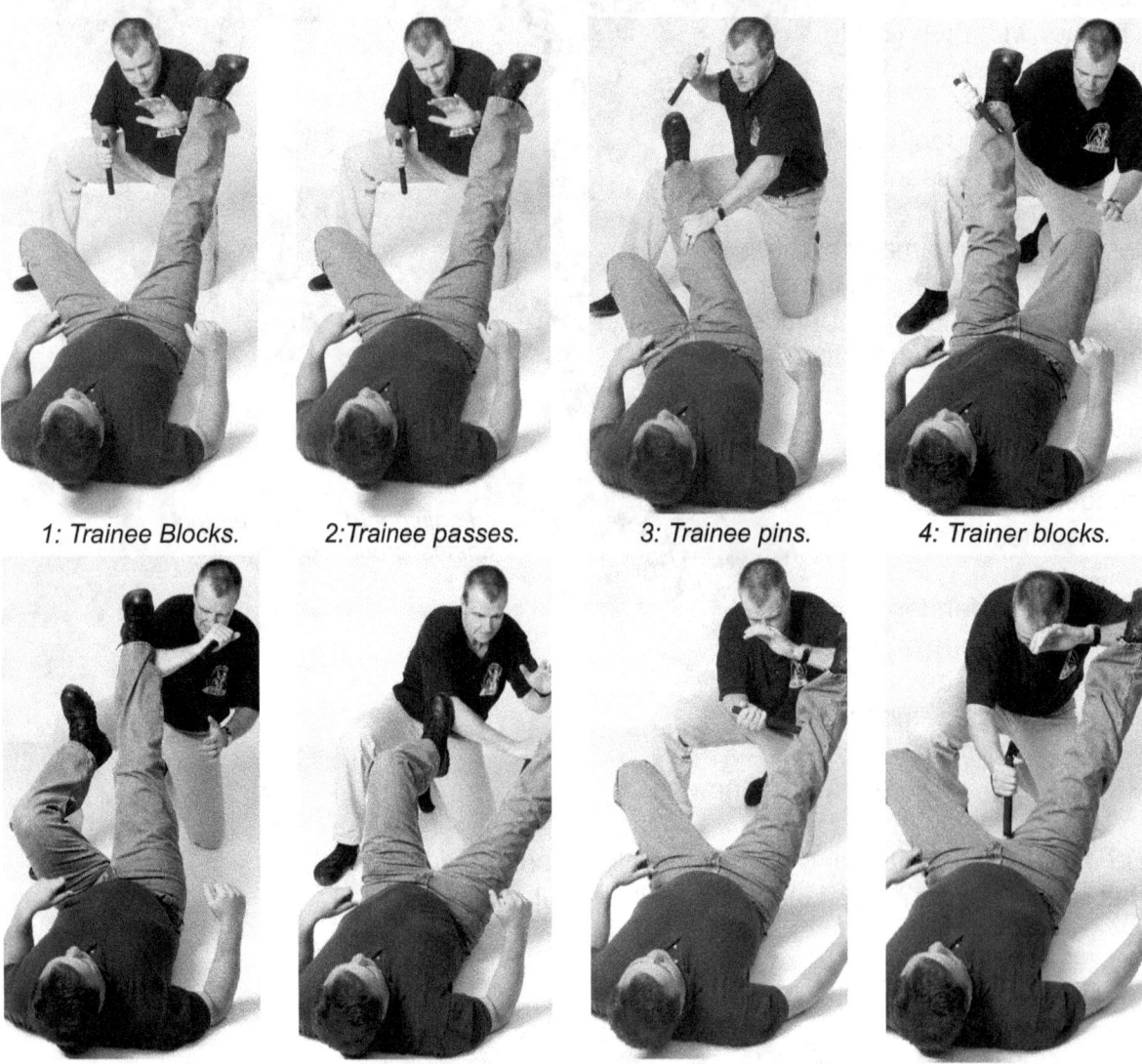

1: Trainee Blocks.   2: Trainee passes.   3: Trainee pins.   4: Trainer blocks.

5: Trainer passes.   6: Trainer pins.   Half-beat slash example.   Half-beat stab example.

1/2 Beat Inserts and Interruption Examples

* half-beat saber slashs to lower leg, upper leg, groin.
* half-beat saber grip stabs.
* continue to invent half-beat leg options.
* half-beat ice pick slashes to lower leg upper leg, groin.
* half-beat ice pick grip stabs.

Rattlesnake Top and Bottom-side Situations
1: He is on top.
2: He is on the bottom.
3: He is on your right side.
4: He is on you left side.
5: He is attacking unarmed or armed.
6: You are kneeling or seated in a chair or car.

Side-By-Side Combat Scenario Situations
SBS Clutches 1: High in the death grip of.
SBS Clutches 2: Low in the death grip of.
SBS Clutches 3: Split high and low in the death grip of.
SBS Clutches 4: Righty vs. lefty grips.
SBS Clutches 5: Arm wraps and clinches.
SBS Overview 6: Review the whole "In The Death Grip Of" section for details.

Side-By-Side Combat Scenario Solutions
Solution 1: Kicking the enemy in the groin, stomach, chest or face.
Solution 2: Oral attacks! Biting, yelling and spitting.
Solution 3: Limited use of the head butt. Use as a last resort.
Solution 4: Knife attacks to defending limb and attacking limb.
Solution 5: Circular releases.
Solution 6: Violent yank-always and pull-outs.
Solution 7: Crawls and scissor leg entries and escapes.
Solution 8: Unique problem-solving for specific catches. These are best expressed through predicaments found in freestyle, chaotic combat scenario practice.
Solution 9: Rattlesnake Releasing Techniques. Whether topside, bottom side or side-by-side, follow many of the same tactics used in the standing *"In the Death Grip Of"* releases.

Jim McCann at the original UFC gym demonstrating the dangers of wrestling muscle memory versus weapons. This triangle lock/choke allows the knifer to simply switch hands!

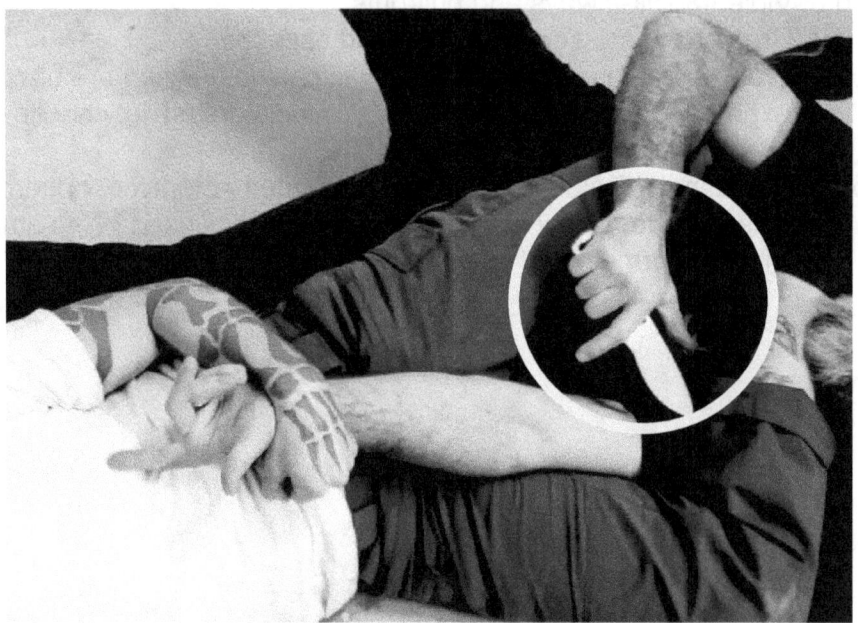

I did my best to try and hide their identities from embaressment. The stupid knife grips are all full time stupid, but worse is doing them in groundfighting where any movement can lead to losing the knife/disarming, or lame, weak applications.

FORCE NECESSARY: KNIFE!
KNIFE FIGHTING
Knife vs. Hand - Knife vs. Stick - Knife vs. Knife - Knife vs. Gun

Addendum to the Knife

"I have been writing, training and learning for years since the first edition of this textbook in 2009 was published. I have improved the previous pages with new information, and here are some personal observations on the subject I have written since."

– Hock

Chapter 30: Addendum to the Knife

Addendum 1: Making a War Post

Nothing has improved my stick or knife training more than hitting hard targets with real weapons. Chop down these war posts! You can afford to replace them cheaply and quickly. I made two posts for about $80 to $100 through the years, however in the 21st Century it might cost a bit more.

1: To build a single war post, you need -
   a) a cheap metal bucket.
   b) a cinder block.
   c) a fence post.
   d) a bag of cement.

2: Cut the fence post down to your required height/size.

3: Mix the cement into the bucket. Set the cinder block inside, holes up. Fit the pots (or, in this case- posts, into the holes, and put cement into the holes. Leave the post to dry. Support them so they will stand straight.

4: Here are single and double dried-and-set samples.

5: Attach add-on arms with some rope. Watch out for splinters!

Addendum 2: Karambits – The Karambit Gambit

There's an old story going around about me and a karambit. The tale goes that during a seminar, lunch break, in the 1990s, a guy walked up to me and showed me his karambit, and I looked at it, opened a nearby window and threw it out the window of a two-story building. This isn't true. I would never do that to a guy's property. I can say confidently that not only would I not be so rude as to throw his property out a 2-story window, I would never throw my own karambit out a window either – because I would never own one in the first place.

Being somewhat in the business of knives, I am all too often shown karambits and asked questions about karambits. You know, the curved bladed knife that looks like a single animal claw. Some folks think they are God's gifts to knives. And I am shown and see way too many karambits. I see photos and photos of them on the web. God, they look cool. All kinda science-fictiony. Klingon-like. Deadly. Tiger-paw looking. I can honestly proclaim I have never seen a karambit I didn't think was very, cool looking. Lord knows I don't want to be attacked by one. But I don't want one. Don't need one. Because of the Karambit Handicap. I hope I can Iserve as a source for people with these questions for me and questions in general about the true value of the knife in the big picture. I hope I can offer some reasoning and answers about the subject. The following are my personal beliefs and how I have come about them. If you love you some karambits? That's fine. Enjoy a happy, healthy life. For me? Out the window they go! Figuratively of course.

As a questioner, as a skeptic, never a fan-boy, not naïve, I just don't fall for people or systems. Worshipping a system-head or a system is a recipe for potential mistakes and failure. If you never question your revered leader, you fail to evolve. So does he and the system he does. Or folks never question gear of the revered. Do you think you must fight with a Klingon knife because you worship the culture, look, feel and history of Klingons? Or are you really looking to fight and survive with the best edged weapon? Are you so mystified by a culture that you can't see the faults? I know Systema people who like it so much, they start believing in and supporting Communism. I know Kung Fu people who change their religion. Communism and Zen Buddhism should have nothing to do with kicking a guy in the nuts or selecting the best knife. If you want to learn how to fight with hands, sticks, knives and guns? Keep hero-worship OUT of the picture. Keep system worship out of the picture. I think this imperative. I constantly see folks doing unnecessary things just because Dijon Superfly does them, and they are too blind to question. I think you can respect a system head and, or a system, but worship is not good. How much do you salivate?

The Karambit Handicap. I cringe every time I see an attendee with a karambit trainer in a seminar. I know that this person will have an extra and hard time doing even the most simple, obvious, historically successful knife moves. My knife training course is built to be as simple as possible, as fast and effective, with the obvious and simple tools, which are the straight blades. Curved blades complicate simplicity.

Am I just untrained and dumb in the wild and wooly ways of the karambit? I frequently get hate mail over this from fan-boys and faddists, people apparently in some sort of odd, over-love with their knife. Someone will always suggest that I am ignorant and suggest that maybe I should take a karambit course and see the wonders and magic of the knife. Dear Dipshits, I was force-fed balisong and karambit material since the late 1980s, since before many of you reading this were born, or as they say, were mastering potty training. Force-fed in multiple training trips to Negros Island and Manila, the Philippines, and many times since there and here since. These knives were part of curriculum we had to learn all the way to Filipino black belt, along with a lot more of straight knife material. I will always prefer the straight knife to the karambit, and well - just forget about the odd, opening process with the balisong. I mean, seriously, why bother? (Unless of course you are a weapons, historian of some sort. I am not.). As soon as I held a karambit in my hand, it felt wrong and much of what they asked me to do was clearly unnecessary when compared to all the other straight blade training. As a former Army and Texas cop and an investigator most of my adult life, from arrests, cases and forensic training, I learned the straight knife is far superior and can do everything better and simpler than any curved knife, just about any time. The curve of the blade is a handicap. The more the curve, the more the handicap.

I recall the first time it happened in a New England seminar in the 1990s. A rather famous, Silat guy showed up with his curved plastic trainer. He had difficulty doing even the most simple, primitive knife things all day long. He couldn't stab deep which is forensically the most successful, quicker kill method. It was plain to see that when slashing, his curve and tip would get stuck in body parts. Did he know he had to improvise and construct more steps, more "work-arounds," to get the job done? I don't know because he just flow drilled around the reality like there were no obstacles. Some do see this truth. Through the years the curved blade trainees still appear in my classes. The curve group often has to pow-wow off in the corner to make a simple thing work, because they are mentally and physically confined from the shape of their knife. Their adaptations always involve extra work-arounds and extra training and extra movement to do something otherwise done simpler with the straight blade.

What do I mean by simple, proven moves? One simple example? Studies by the Marines in 1980s – while researching World war II knife tactics in the South Pacific, the USMC study group discovered that the uppercut stab to the groin/intestines, and, or the diaphragm/heart and, or even up inside the jawbone – the common hooking uppercut was a very successful. Successful, but oddly, not really emphasized and in most cases not taught. Yet, Marines instinctively still did them. Naturally. Natural. This research led to the implementation of these very natural moves in training courses. Instinctive. Natural. Simple. Now, can you do this natural, saber grip uppercut into these areas with a karambit. No. You can't plummet a karmabit, even one with a bottom side out grip, as deep and powerful into these vital parts as a saber, straight knife. Aside from results, the saber, straight knife movement is more natural, and the karambit will require extra training and still won't garner the same success. Don't get me started on all these examples as this will become a book and not an essay.

Now look, you can cut somebody with a torn-open, tin can. I also don't want to be attacked by a torn tin can or anything sharp. Broken glass bottle. Nope. A spear? Hell no.

But the question remains is, yes, a tin can will cut you, but is it the smartest thing to use? Do we need the Tactical Tin-Can course? No. You just get a knife. Get the best knife. A straight knife that stabs with deep efficiency potential and slashes without getting stuck in bodies and some clothing and can also, easily perform dozens of life-saving and survival chores.

Sellers of karambits have much sales-pitch, yadda-yadda about the cancer-curing perfections/wonders of the curved shape. They proclaim that just about everyone on the planet already uses, benefits and really needs the really curved knife. EVERYONE uses and loves the karambit, everyone except the real people you see, you know, work with and read about and watch in documentaries, etc. I suggest you challenge every line of the sales pitch because in the end, it is not the selection of the practical.

In actuality...
Butchers don't use them.
Surgeons don't use them.
Cooks don't use them.
Hunters don't use them.
Fishermen don't use them.
Soldiers & Marines in the know don't use them.
People don't use them to camp.
Workers with real labor jobs won't use them.
People don't eat with them (this is a big point).
Prosecutors and police love to see you use them.

If they are so perfect and superior, why are they not used by all humanity most of the time? Try giving a farmer, a factory worker or a camper just a karambit and see how long that idea lasts before they trade out for a straight blade. Give a carpet layer a karambit and he will quickly resort back to his carpet knife. Many, if not most, of the big name karambit twirlers have never been in the military. They just don't know that a military knife in the field must be very versatile and able to perform many everyday chores, as well as possible fighting. (And by using them, I mean predominantly use them. I am sure in my incoming hate mail over this, someone will name a special circumstance where someone drops his regular straight knife and reaches for a curvy hook knife to catch an oddball body, fish or animal or autopsy part.)

The biggest point in the above list, to me is that the human race has evolved to hunt, grow, prepare food and eat with a straight knife. Ever try to eat a steak with a karambit? Cut and butter bread? I have a friend who likes to tease me on this point and threatens to send me a video of him eating a steak with his curvy karambit. I'll bet he can! I'll also bet he can eat a steak with a torn, tin can. The point is, not that you can or can't, but rather - what is the smartest tool to use. And we can't forget, kitchen cutlery has reeked international havoc in self-defense, crime and war. In civilized countries over 99% of all knife violence is with simple, kitchen cutlery. A pretty good success rate for the straight blade.

Chopping off limbs with the karambit. Did we mention butchers above? A good friend of mine, consumed by all things "distant" and Eastern, Oriental and Indonesian, was telling me that a butcher he knew, using a very stout, big karambit with a sharp outside edge, could flip/spin the curved knife and chop off the limbs of large animals in his shop. It took some practice, but he could. The message for me was that the karambit could, if worked right, with the right momentum, chop off big things in a power spin. CHOP! I just nodded my head. Whatever. But such takes more work, awkward applications, etc. and stouter karambits with a sharp outside edge. If it were a big folder? How do you have a sharp, outside edge and carry it? Not in a pocket, but in a sheath...in case you know...you have to lop off a hand. I am quite sure the butchers of the world will still prefer regular straight knives and clevers for more efficient, consistent success. What will be this butcher's tool of day-to-day preference. The easy one. And then I must ask, will you always carry around this oversized karambit with the complete outer side sharp? Whose forearm do you imagine you will be cutting off in your day-to-day? In YOUR world? Jaime Lannisters?

And needless to add, take a guy with a straight, blade knife in a saber grip versus a guy with karambit and let them duel. Who do you think has the advantage? Spar it out. Take two Superflys and spar this straight vs. curved karambit. I can tell you from doing that for decades and organizing/ref experience that the saber grip straight blade has the advantage. Not that dueling is the end-all knife encounter, a final judge, oh no, but dueling can and does happen. And listen to this – this is telling – even the Superflies still teach and use a whole lot of straight knives too. Most teach more straight knife than curved knife. Why bother? If the karambit was God's gift for knife work? Wouldn't they give up on straight blade material all together?

But they look cool, so Klingon and purity'! And Dijon Juan Superfly is soooo cool with his flow drills on youtube!

"Oh my Dijon! Oh my....and…and Dijon does so many arm manipulations."

Do you think you will really hook and push around so many angry, adrenalized arms with a karambit as Dijon Superfly does in a cooperative flow drill on YouTube? And by the way, a straight knife can push arms around too.

Back to Spinning the Karambit. The ring in the handle alone does not a karambit make. I have seen some folks calling a straight knife with a ring in the handle a karambit, just because of the ring. No. It has to have a curved blade to be one. Now, to what degree of a curve, I can't precisely say. I think you know one when you see one. The ring is for mostly for retention and...spinning.

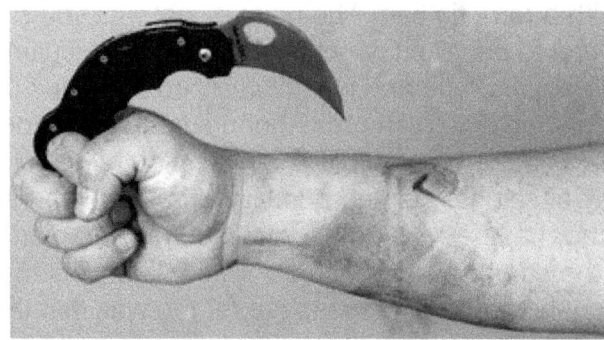

On spinning, another dubious karambit characteristic- the karmabit sellers page says,

"Karambit spinning is showy, flashy and useless without significant training, practice and understanding of the application. New users should not spin karambits until they're intimately familiar with their blade, its balance, the way it fits into their hand in various grips and while in motion AND, most importantly, until they've received instruction." Further, he adds ***"...many people don't use the smaller muscles in the hands and it takes time to build them up."***

Confessions from a top karambit salesman! And there you have it from the source. More stuff to do. More muscles to build. More unnecessary stuff to do.

Straight, bent, curved. The curve of the knife. The more curve, the worse. There are knives on the market that have some bend to them, some just a slight bend, bended/angled with no curvey claw. Some right-angle bends remind me somewhat of carpet knives. The sharp, 90 degree bend of the carpet knife, its position to the handle, is superior to the more curved karambits, otherwise thousands of carpet layers would have invented karambits or they would all use karambits. They don't. Some folks, like carpet folks, work projects that require that sharp point, at the maximum position of their hand grip for the job. As a detective I have worked some serious assaults involving carpet knife attacks.

Stress Quick Draws Issues. A comprehensive knife program covers stress quick draws. It seems all modern knives now try to have some pocket catching device that facilitates a quick folder opening. But some don't. Sometimes people get their folder out but in the heat of the fight, can't open right away. The folder then becomes a palm stick until its opened. The selected knife when folded should protrude from the top and bottom of the fisted hand, and it should support the hand inside the fist for punching. I have a pretty big hand and have tried punching heavy bags with various karambits. Due to the curved blade, the folded knives are quite wide and they all hurt to punch with. Probably I might find one not as wide someday, but with all the other negatives surrounding the karambit? I don't go about searching for it. But this wideness when punching is another survival reason/problem to avoid the karambit.

Even the Wolverine's claws are rather straight!

One of the great advantages of the reverse or ice pick grip of a straight blade is it's ever-so-natural, stab application. There seems to be an inert, intuitive hammer fist application with a reverse grip stab. Think of the power of just a hammer fist. It alone breaks many boards, many ice blocks, many pieces of cement. Imagine that force delivering a straight knife stab! But wait! Now hold a karambit in its reverse grip application, as in the curved end looping out of the bottom of the hand. Gone is all the hammer fist intuition. Gone is the simple, practical, stab and its extra power shot potential.

The somewhat bent edged weapons list might include the infamous kukri. The kukri is not a curvy karambit. It has its own heft and is used much like a straight edged weapon. Straighter? "Benter?" Curved? These bended ones are better than the curvy ones, and seem to have some "hammerfist-like" and "punching-like," natural applications. But, the more the bend? The more the pointy curve? The more problems. To use them as efficiently as a straight knife, which cannot be completely done, you have to add-on, learn more, have extra tricks to stab and slash. And, speaking of hammer-fists, the hammer fist is a very natural movement, with very natural target acquisition, and really supports the reverse or ice-pick grip, straight-blade stab. Why ruin that principle with a curved blade that sticks out and then forward from the bottom of hand, killing the hammer-fist instinct. So...more karambit training is therefore needed. More extra training.

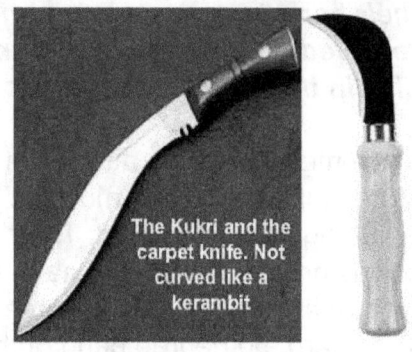

The Kukri and the carpet knife. Not curved like a kerambit

As mentioned it gets stuck in people and things. The curved point is called a hook, because...it hooks. I see the karambit practitioners simulating cuts with figure 8 patterns and X patterns in the air, or in front of partners. No contact. Do they not realize that with contact, their point embeds into the person and the bones and the clothing, gear, etc? X pattern over. Figure 8 pattern over. And now they must learn an extraction technique, unique to that knife. Extra stuff to learn. (this is also true with the tomahawk/axe craze. On first impact? THUNK! NO more slap-dash, dancey, prancy axe moves, just a big-ass axe sunk into a skull or chest. Extraction! Use foot if needed to push-pull)   The most

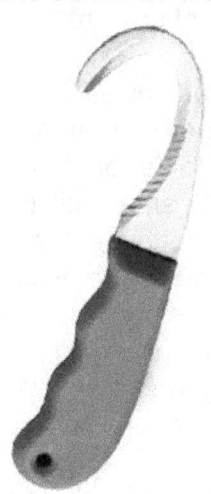

curved knife "out there?" The classic rescue knife. One carried just to cut seat belts and ropes. You can't even clean your nails with this one. I think that anyone can see this is really limited in overall use. The more the curve, the less you can do. I am sure when you need a seat belt cutter? You really need a seat belt cutter. So, get one and cram it on your belt. Squeeze in that two inches more next to your shark repellent, and radiation pills, for those times you really need stuff like that. (Oh, and yes, a "regular" knife can cut a seat belt too, and I'll bet has many more times than a specialized seat cutter.)

And lastly, need we discuss the stigma again of this Klingon-looking knife. It is bad enough to use any knife for self defense, but this knife, by its very appearance also causes negative, legal prejudices to the police, the prosecutors, the courts and to juries. Think of it in terms of pistols. Would you rather defend yourself with the "Widowmaker" pistol? Or..the "Peacemaker" pistol?" Yes, these...things...count. In a recent self defense courtroom trial, Assistant District Attorney in Texas Aaron Bundoc also said of a defendant's self defense use of the karambit,

*"It was not a self-defense tool as Hernandez alleged." He said
"...a karambit is a combat weapon designed to gut and butcher people."*

Just ONE example. Look, what do I care about people, their fixations, fascinations and hobbies? Why should you care what I think? Some people love history and weapons. Some people like to crack bull whips, while the whip is on fire! Get a hobby! Get a karambit and mess with it. Do all that extra training. Place it on a rotating pedestal in your den. One in each pocket and on a neck chain. Get the t-shirt and ball cap. Follow the Dijon. Smile. Live long and prosper. These are just my personal beliefs and opinions. I know I never want to be attacked or cut by a karambit, or a torn, tin can or a rescue knife. Hell, I hate paper cuts. But to me, a karambit is a handicap to sheer simplicity and ultimate practicality. People are just too damn hypnotized by the shape, culture, history, hero-worship and system-worship.

What did they tell us in school years ago, when writing an essay? "Contrast and compare." If you really contrast and compare, without bias and fixations, fandom and fads? What do you come up with? I will never bother with, or waste my time teaching, a karambit course. Do please, however continue to show me your karambits. They are all very cool looking. And I certainly will not throw them out of any window. Only, you know,...figuratively.

History tells us that many curved swords were used in horsemen versus horsemen, or ground troops versus horsemen battles. But, the OUTSIDE of the curve was used. They feared that straight swords or the inside of the curved swords and their tips might stab into and catch the enemy, causing them to lose their swords from the force of a horseman.

Addendum 3: I've Heard the Whisper. Have You? More on the Secret Tip of the Reverse Grip "Insider" Tip.

"If you see a guy hold a knife like this (reverse grip) watch out! He really knows what he is doing."

Said the tipster who knows nothing, anyway. A receiver from another tipster who knew nothing. I have heard this "insider tip" numerous times. One time, I was even told this tip by a guy who had never been in the Marines, but heard this tip from a "Marine friend." The irony was I was in A Triangle, VA, hotel restaurant next to the Marine base Quantico, there teaching Marines among other things – "knife." Some of it saber grip knife.

I have nothing against the reverse grip. I teach it too. But, how many times have you heard or been told, or been taught the reverse grip is the only grip to use? Because there are some popular courses out there preaching this idea. I am often both depressed and fascinated by the paths and ideas of various knife courses. Some appear to me like death cults, others have what seems to be oddball, incomplete conclusions.

Take for one example, the obsession with dueling. While dueling can happen, it is not the only thing that happens in a knife fight. (Ohhh, like ground fighting?) Most never cover legal issues and just cut and stab away. Some way over-emphasize Sumbrada patterns. I could go on. Another is the various obsessions with knife grips.

In the 1980s, I was at a Dan Inosanto seminar and he said,

"There is no one perfect knife grip, just the best one for the moment."

Man! That little phrase stuck with me forever as a baseline. I worry about knife grips too. With my Force Necessary: Knife course, I insist that a practitioner learn and experience BOTH the saber and reverse grips. Then after much COMPLETE study, they choose what their favorite grip is, based on the "who, what, where, when, how and why" of their lives. I never tell them to automatically favor any one grip. Pick one, but make it an EDUCATED choice (and not a whisper tip choice from the ill-informed). I like both grips. But some don't like or respect both grips and for odd and incomplete reasons.

The first time I received any training with a knife was in a Parker Kenpo class in about 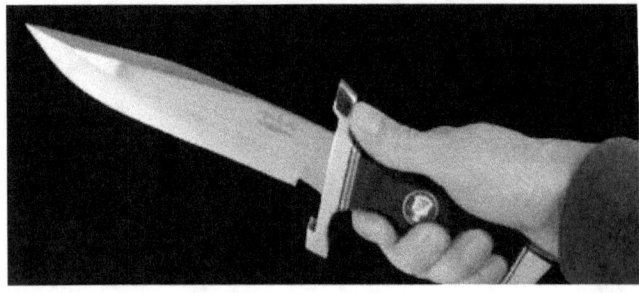 1972 when they received Kenpo knives and dull trainers from Gil Hibben. The knife was designed to be held in a reverse grip and used in a kick boxing format. This initiation caused decades of Kenpo-ists to hold their knives in a reverse grip as though Moses, not Gil Hibben, had handed them knives from the mountaintop. In the late 1990s I was teaching in a multi-instructor camp and a old Parker Kenpo black belt was there with his little group. I was covering saber grip material (a pox!) and he would hurd his people over into a corner and show them the "proper, reverse grip way" to fix what I was doing saber style. Which by the way his was more complicated and somewhat more awkward than the simple saber material I was covering. Finally he had to approach me and said,

"You know, the reverse grip is the superior grip."

I said, "No it isn't."

He glared at me.

"You can do just as much or even more with a saber grip," I added. "Actually a few things more."

He stormed off – his 7th dan "master" self – all upset that someone younger in blue jeans and a polo shirt could tell him flat-out, "No". (Oh, did I mention that he also makes and sells reverse-grip-only knives? To go with a reverse grip only training course?) My next session, by the way, I covered some reverse grip material, as I am…an equal opportunity stabber. Reverse grip was just not on my planned schedule for that time.

People pick grips for odd reasons too. I recently read a note from a guy writing about his grip choice. He said he once saw someone with a training knife, saber grip, stab a mitt and the guy's hand slipped up on the blade. This made him pick the reverse grip. Really? That? Well, as a detective who has investigated knife crime for decades, I can tell you such slippage happens with BOTH saber AND reverse grips. It depends on the knife! It's shapeness, handle texture and any guard/hilts, design, etc matter. Dear Reverser – your hand slips down on a reverse grip too. Jeez. In fact, a hammering reverse grip deals a mighty blow, and your hand needs to be protected perhaps even more, versus an ice pick grip slip.

There are people and courses out there now that mandate the reverse grip, ignoring the saber grip, therefore treating the saber grip as worthless? Mindless followers argue that:

"Well, the reverse grip is the Filipino way."

"The reverse grip is the Pekiti Tirsia way."

"Ex-super-copper Pete Smith says the reverse grip is the best."

People! People….People. FMA does BOTH grips. Pekiti Tirsia also teaches tons of saber grip material. And Pete Smith? He is just, flat-out, wrong-headed about this. He is not as smart as you think. He is not as smart as he thinks.

And I will go you one more reverse grip oddity that's worse. There is a sub-culture out there that wants you to fight reverse grip, with a single edge knife, with the one sharp side, "inside," facing back at your body. Not edge out to the outside world, where the enemy bodies are attacking you, but edge back at you. Dull edge at the enemy. This is its own unique thinking disorder. The world you are fighting is OUT there, not in your armpit. And why must you hook and haul ALL your enemies, ALL the time inside and close to you? Can you? Can you hook ALL people, ALL the time, in for your one- trick, special magic?

You should have a double-edged knife, which is legal in some places, or you should have a knife with one complete sharp edge and the other side/edge 1/4, ½ or ¾ sharp anyway. (the so-called "false edge," a confusing term for "civilians.") Then the edge-in-only argument becomes moot, because you can still slash and stab, and stab and slash the outside world. But to purposely plan and buy and limit yourself to one edge, then limit yourself with that edge-in only, is…is…just a thinking disorder. It's a dangerous, thoughtless, thinking disorder. People even make up these excuses.

"Well this type of knife is meant for…"
"I like this type of knife because when the guy is here, I can…."
"It works really good for getting his arm away when he's choking me…"

Dude, your "meant for" list is too small when you consider all the possibilities you might face and with each specialized shaped knife in reverse (or saber) grip. So then what? Are you going to carry 3 or 4 different "meant-for" knives for when the he's "here," one for one he's "there," and two for when he's everywhere? To cover all your "meant-fors?" All knives work well on an arm when someone is choking you. Dude, maximize your survival with the most versatile knife. "Meant-for knives" is like putting 2 rounds in a six gun, and then carrying 3 pistols.

Why in the world…WHY…buy a fighting, survival knife to save your life and the lives of others that is single-edged? And then…and then…carry it, and use it, edge inward?

You do know sometimes in a fight, things don't always go as "stabbingly" as planned. The reverse grip, edge-OUT offers a slashing possibility, an ADVANTAGE that might diminish the opponent. A double-edged, or 1/4 or 1/2 edge or 3/4s on the other side is ALWAYS going to be an advantage over a single edge knife.

Reverse grip tip in accidents. When the Samurai commit suicide – seppuku – they do it with a reverse grip. There's a reason for this. It's easier. The tip is already aimed inward. (Incidentally, the helper that will kill him if he fails? Holds a saber grip.) This "easy-tip-inward" is one more point no one seems to consider when raving about reverse grips. The tip is often aimed back at you. About once every 4 weeks on the nightly reruns of the "Cops" TV show you can see a reverse grip knifer get tackled by cops. When they turn the suspect over? He was – "self-stabbed." One of our lesson plans with the reverse grip is self-awareness of these maneuvering problems.

I cover dropping to the ground in training and we see people stab their thighs with training knives, and when tackled or shoved against the wall, we see the "Cops" TV incident of the accidental self-stabbing.

Tell people to practice falling while holding a knife and watch the accidents. If people are running and holding a reverse grip knife, and they trip and fall? Watch out! There are more reverse grip "selfie" accidents than saber grip ones. Just be aware of this.

There are many pros and cons for each grip. I HAVE NOTHING AGAINST THE REVERSE GRIP. I teach it also, and I believe I am better and more thorough and thoughtful than most…or I wouldn't try. I just don't like the blind acceptance, the secret whisper about which grip is the best and which is to be ignored. And folks, the single-edge-inward version is just dangerous nonsense. And It bothers me that people thoughtlessly accept courses about these things. Question everything. Get educated. Then pick a knife and a grip you need for your world.

The knife is a very forgiving weapon, in that you can do a whole lot of screwed-up, stupid shit with a knife and get away with it. This is not an excuse to stay stupid. You have to think beyond that to create a comprehensive program and promote real knowledge.

Look, I really don't care what grip you use as long as it is an EDUCATED, informed choice, and not some mindless, mandate from some short-sighted, and, or brainwashed person, or from that ignorant knife whisperer who knows somebody, who knew somebody else, who...

Having written my extensive and popular knife book, which took years, studying crime, war and forensic medicine, working cases and the streets, you discover that there are about some…what.. 30 or so ways to fight people with a saber grip and about 25 ways with a reverse group. But, after 8 ways each, does it matter?

Some responses:

"Years ago I was grappling with my training partner and had a wooden training knife in the reverse grip. Long story short, we went to the ground, and I fell on the training knife and caught the tip in my ribs. That definitely made my eyes water. Since then I'm very iffy whether I'd ever use that grip with a live blade in a real situation. I still train both grips, but I much prefer the saber grip due to the added reach, maneuverability, and the sharp end isn't pointing back at me most of the time." — Neil Ferguson, USA

"I was at a seminar where an "expert" told us that, 'If we see a man using a knife in that reverse grip to run! That man is a professional!' First off, I'm gonna run either way genius! A 12-year-old can get my wallet with a two inch folding knife. Stitches are expensive! Furthermore most women use that grip when they stab. Are they professionals?" — Zane Issacs, USA

Addendum 4: Personalties of the Knife

Knives have personalities. The generic look. The generic history. Slashing look. Stabbers. The personal attachment look. What is the personality or your knife? I think there are several factors in knife personalities.

Culture of the Knife Personality

One is the culture of the knife. Certain edged weapons have a history, a geographic flavor. Just think of the Japanese Tanto. The Kris. The Bowie knife. The Italian stiletto. The Medieval dagger. The double-edged, commando knife. One in the martial business, or the knife aficionados, or makers recognize the aura/genre of many knives. This cultural attraction alone might be a main reason someone buys to collect, or buys to carry a knife. Somehow, some way, the look captures one's fancy, imagination, expectation or whatever connection to books, movies, TV or past affiliation. Sort of a mysticism we mentally project upon a simple inanimate knife. After all, what makes us select the cars, pants, churches or sports teams we do? We are tribal, particular and peculiar from our hats down to our shoes. Hats and shoes as in style that is, not in size. We can't change the size of our head or our feet. We can change the size and shape of a knife, but will the size be appropriate for our…"heads" and "feet?"

Slashing, Hacking and Stabbing Personalities

The shape and size of the knife tells an experienced handler what it can do best. Some are better hackers. Some long, thin ones are better stabbers. Some are wide and are better shaped for slicing. Like a carpet layer needs a certain angle for exactly what is needed, so do all knife users. A novice to so-called, knife "fighting," a new-be to say, construction work, will not know what kind of knife does what best. Experience and education is called for.

Personal, Knife Personality Examples

I knew a Green Beret, Vietnam vet who passed on standard Army/government issue knives and preferred his old own Bowie Knife, replete with a carved stag handle. It was a family heirloom you might say, and therefore more important to him than any generic, legend of Jim Bowie. He said it gave him a certain power, a certain mojo from which he garnered mental and physical strength. This is a personal touchstone, reminiscent of many cultures, such as some of the native Americans might carry a medicine bag of mojo. Same-same.

Another friend of mind sought an old-fashioned, traditional looking (and hard to open) pocket, folding knife with stag handles, with multiple blades, because his dad had a similar one and it was lost through time. Both, more "personal, private" personality, touchstone selections. Still, with game points awarded for symbolic and personal mojo, on the battlefield or for back porch whittling, the knife size, shape and handle must fall within a scope and range of usable practicality and common sense. Switch this over to a parallel concept – you wouldn't a pack a flintlock pistol around for self defense, just because you love the early American history era. Extrapolate this idea over to other weapons and survival.

What personality knife do you really need? – Not just want for whatever abstract reason, but need? I think we have to return to the classic, Who, What, Where, When, How and Why questions I use all the time to best determine this. Let's review the Ws and H questions.

- Who are you to need a knife?
- Who are you to carry a knife?
- What do you really need or want a knife exactly for?
- What do you exactly expect to do with this knife?
- What training do you have to make this a wise choice?
- What are the local laws for such a knife? What state and, or country do you live in?
- What happens next? You use the knife and what will the police and prosecutors think of the name and look of your knife?
- Where will you carry this knife? Job? Protection? Handiness?
- Where on your body will you clip, or sheath or cart your knife?
- When will you need this knife? Work time? Off-time? Daytime? Nighttime?
- How will you acquire this knife?
- How will you use it? Do you know how?
- Why will you select a specific knife?

Another, longer "what" question. The chicken or the egg? What came first for you? Or, what will come first, if you are just now thinking about knives? That mysterious adulation of ..."the knife," and then a knife training course? Or did you need a knife first for a task first, then seek a training course? This consideration might help clear a path for your knife selection and proper training. The collector, the historian, the practical user, the adulator? Who are you?

But that last line of questioning…the "why." Why will you select a specific knife? I suggest that you do not make a selection based on looks, genres, eras and or culture alone. I think you should select a knife on its ultimate practicality. Of course if you are a collector looking for this or that showpiece – "I own one! It's a beauty!" – have fun! (I am not much of a collector of things so I cannot relate to this, but of course, I do understand the hobby.) Or, if you are fanatic about say, old European sword and dagger fighting. Whatever. Get those weapons and mess around with them. Have fun and exercise. Shoot flintlock firearms (just don't carry them as a self defense weapon).

Knives have personalities. The generic look. The generic history. The personal attachment. If you plan to actually carry and use a knife? Whether on the job as a telephone lineman, a surgeon, a soldier or a cop, or just a citizen with a hankering for a knife, think of them as tools and well…think of them as shoes. You'll be wearing them too, and like your hat and shoes, you can change the style, but you can't change the size of your head and feet. Get the appropriate tool/knife. See clearly, be fleet of foot for the trails and paths of life, Kemosabe. Don't stumble around with the wrong size, else you'll trip, fall and fail. And like "running with scissors," running with the wrong knife can be a minor or costly mistake.

Addendum 5: Fencing and Knife Fighting?

> Sword.
> Small sword.
> Big knife.
> Small knife.
> Sword fighting.
> Epee sport fencing.
> Dueling.
> Olympic fencing.
> Knife fighting.

Enthusiasts like to discuss and compare dueling and fencing with knife fighting and big knives and swords and smaller knives…and, and, and… The topic comes up now and then on how fencing skills help knife fighting skills and also, oddly, how fencing skills helps fighting in general. For me? Not so much.

It is confusing to discuss these things unless you set down some edged weapon world definitions. Some people can't adequately define them, which causes a confusing debate and conclusion. They might say, "the best knife fighting training is fencing." Well, what kind of fencing? There are several kinds of fencing with small letters and capital letters. And several different tools used when fencing.

A sword is…well, a sword. There are all kinds of swords, you know. We immediately think of the olden days and the swords of knights, and Cossacks and the Three Musketeers. As defined in most universes, "A weapon having various forms but consisting typically of a long, straight or slightly curved blade, sharp-edged on one or both sides, with one end pointed and the other fixed in a hilt or handle." A key word in our comparison is "long." We might add that some swords are best wielded by two hands. Many still by one hand.

A Big Knife is…well, a big knife. It certainly will be handled by a single hand. Two hands need not apply. Some people consider any knife with a blade over 6 inches and "less than a short sword" is a big knife. I recall one knife enthusiast remarked, "…is defined by a culmination of its features, a critical mass of its qualities." That has a ring to it.

People think of the ubiquitous Bowie Knife as a big knife. It is also knife-lore that a small knife is concealable.

A Small Knife is…well, smaller. One-handed. Easy to conceal, as the topic of concealment often comes up with defining sizes. It would stand to reason that a knife with a blade under 6 inches could be called a small knife. It is also knife-lore that a small knife is concealable.

Dueling is…well, you know, right? Look up "duel," and you will find terms, like "pre-arranged combat," "observed by witnesses," "with guns and swords" and other uses like a "duel of wits." Today, when we think of edged-weapon fights we often just think of a knife versus knife fight, sort of a touch-and-go, kind of deadly chess match, tag kind of thing? You'll often hear me talk about the "myth of the duel," in that unarmed or armed fighters "on the street," usually aren't in perfect stances and pre-arranged distances, in and out, experimenting with jab results, to "win in round three." Two fighters crash! But two fighters can indeed break apart and this might look like a movie duel for a period of time. A bad movie, but sure, which is why to be comprehensive, we must train for these bits of dueling times too, but the fight is much more.

A Knife Fight is…uglier? Less classy? And how is it that two guys are fighting with knives anyway? Who, what, where, when, how and why? To enter this ugly realm, I often say, "knife fighting is like football/rugby with a knife." I realize this is a sloppy, somewhat incomplete description, but it makes a point about how messy knife fights with real angry people can get. Then there are those folks that believe ALL knife attacks are hidden-knife ambushes. Or ALL knife attacks are madman stitching. Not so. Not so, as a true historian of war and crime will know better about the diversity.

Fencing. When you fence, it can be sword fighting, and "sparring" training with any and all swords. You certainly fence in any sword class. Which next leads us to ponder, Fencing (with the capital "F"), the sport of "Olympic Fencing." It is described as an "art or sport of using a foil, epee, or saber in attack and defense." Most of us envision the single lane sport of forward and back, wispy epees and those special metal helmets and white suits. Most know that the modern fencers use electric epees and suits. But if you are a Renaissance sword person, fencing (with a small "f") just means using regular, dull swords to train and fight with. This is different form the sport of fencing. Is an epee a sword? Epee is French for sword, but today's Olympic Fencing sword it is not like a "sword-sword" the Huns, Vikings or Samurai would use. It's a wispy thing, often with electricity running through it. For a sporty game of tag.

Martial artists reference Olympic Fencing at times, or they think they are when they say "fencing." Some go on and on about how we can learn oh so much from Olympic Fencing. I just don't think so. I don't share the love. Jeet Kune Do practitioners know that Bruce Lee praised and studied Olympic Fencing for the fast lead arm and explosive leg work. At one time he organized boxing, Wing Chun and Olympic Fencing as foundations of JKD. I can't help but think that in the big picture of all that could happen in a standing, seated, ground, hand, stick, knife and gun fight, how Olympic Fencing could be so important. There are only a few, slim tricks that cross over. Any exercise is better than no exercise. And Bruce wasn't worrying about knife fighting when working on boxing and trapping hands, even though Olympic Fencing was about "edged weapons." I don't like too much of a Olympic Fencing influence in knife material for several reasons.

* One is that Olympic Fencing is a suicidal game plan. I sacrifice my position, my… everything just to touch you first. That is all I care about. "First Touch." I touch first, I win. That is the Olympic Fencing, training goal. First tag. No matter that, even if the other "blade" cuts my throat after my "first touch" win. I won! My first touch, sport win may not be a successful real fight ender (what knife attack is, and someone should always prepare for the return attack. Sword fighters get this. Knife and sword fighters know that the enemy gets wounded and still attacks. Olympic Fencers don't care! Their training doctrine lacks steps 2,3,4, and so.

* Another disconnect for me is in Olympic Fencing the training is based on a long weapon which causes a fencer to stand more sideways. The shorter the enemy blade, like with big and small knives, the more a person can turn their body into the fight and utilize the support arm if needed.

* Plus, Olympic Fencing footwork training is just forward and back not side-to-side. Just think of the narrow strip they play on, what they call the piste or strip. Regulations require the piste to be 14 meters long and between 1.5 and 2 meters wide. The last two meters on each end are hash-marked to warn a fencer before he/she backs off the end of the strip, after which is a 1.5 to 2 meter runoff. Errol Flynn and Basel Rathbone would duel everywhere, up and down elaborate stairs, swing on chandeliers, leap across roof tops.

* The helmet. No one likes to have long, sharp things stuck in their face. So in training and competition, folks wear a helmet. The helmet distorts your confidence, The helmet distorts your system doctrine. No training or sports program will be perfect. We do the best we can. We need to be safe, etc. I'm just saying…

I like to watch Olympic Fencing and I can recognize what great Olympic Fencing-centric-only skills they possess. If you are unfortunate enough to get into a knife versus knife fight, somewhere, anywhere, and there is a distance break…and you can't leave?…A few seconds might look a bit like knife dueling and some sort of "fencing." You should train for this possibility with the type/length of knife you will probably use. Reduce the abstract. But I would not rely on Olympic Fencing methods as the best training to save my life in a knife fight, or any fight for that matter.

Addendum 6: **If I Pull my Knife? And He is Carrying a Gun? Will this Cause Him to Pull his Gun Out? Will I Cause the Problem to Escalate?**
So often people want *Magic Bullet* answers to a lot of self-defense questions. There's always big talk in the self-defense industry about "avoidance." If too late to avoid, then next up in the event list is what they call "de-escalation." But avoiding and de-escalating a common knucklehead before a fight starts, or a mugging starts, is now a cottage training industry. Some trainers confidently dole out solutions to confrontations in three to five steps or present mandatory checklists.

> *"Say these things that I tell you!"*
> *"Do this!"*
> *"Do that!"*
> *"Stand like this!"*
> *"Don't ever...."*

Now, I think it is certainly good to be exposed to all these ideas and methods. Sure. Do so. But as an obsessed skeptic, I see the caveats beyond the simple advice. I don't know about certain kinds of solutions, magic bullet words, or stances when confronted or attacked.

I have investigated a whole lot of assaults, aggravated assaults, attempted murders, and murders through the decades; and while there are identifiable patterns and surprises, chaos can sure still reign supreme. But let me summarize by calling it all "situational." In the end, solutions are situational. Like calling plays in a football game, it depends on the situation. How you stand and what you say or do should be situational. Custom-built. (This essay is primarily about pulling out a knife but does and could certainly relate to pulling a pistol, too. It's just that if this was a "pistol-centric" essay, I would be writing more about pistol situations.)

So, there's an argument! Then a fight! Given you have already performed all your popular avoidance and de-escalation steps ... you are armed under your coat or in your pocket with a knife or even a gun, and this verbal stuff just ain't working! The mean man won't leave! Do you pull that knife out? That weapon out? There are some situational concerns with doing this; and these concerns certainly do involve his possible knives and guns and the overall escalating ladder of weaponry, violence, and legal problems.

Here are a few facts and related ideas on the subject to kick around:

Fact: Some people do leave. For many a year a few years back, the Department of Justice reported that 65% to 70% of the time when a knife or pistol is pulled in the US, the criminal leaves you alone. Simple statement. I have often heard the easy average of 67% used. (Sticks, by the way, are not in these study figures.) I must warn folks that this is not as clean and simple an escape as it sounds. There are many emotional, ugly events that happen in this weapon-presentation/confrontation, even if the bad guy does leave. Trauma and drama. We discuss these details in certain topical seminars and other specific essays.

Fact: Some people don't leave. The good news with the 65% to 35% split is you may only have to fight about 30% of the time! So 30% of the time, the opponent does not leave and the fight is on, whether he is unarmed or armed. The bad news is when you are now in that "unlucky 30%" or you might say you are now a 100%-er. You are 100% there and stuck in it. A hand, stick, knife, or gunfight!

Fact: Some people are armed. General stats quoted for many years past say that 40% of the time the people we fight are armed. A few years back the FBI upped that ante to about 90% being armed! A shockingly high number for me to grasp. And another gem to add in is that 40% of the time we fight two or more people. Hmmm. So 40% to 90% armed times 40% multiple opponents. Not a healthy equation. Lots of people. Lots of weapons. Lots of numerical possibilities. The "smart money" always bets that the opponent is armed.

Fact: Times and reasons to pull. Logical and physical. Time and reason might seem the same, but defining times and reasons in your mind and for your training is smart. Time equals "when" and reason equals "why." Two different questions. The motive and the moment to move. Either way, remember there must be some real danger to you and danger to others for you to take weapon action.

The Why? Two Reasons to Pull: There are two reasons to pull your weapon out. The first is to stop violence before it happens. The second is to stop violence while it is happening.

The When? Two Times to Pull: There are two generic times to draw your weapon. The first is when you can predict problems and pull before the incident happens. It's always said that the best quick draw is pulling out your weapon just before you need it. And the second pull is during the incident.

More Facts: Pulling during the incident. I have written and lectured in the past about why people do and do not draw weapons once a physical fight has started. They are in this quick review:

1: He is carrying but does not draw because he actually forgets he is armed. Oh, yes, this happens.
2: He is carrying but does not draw because he is smart enough to know that this incident does not deserve the legal and physical consequences of pulling a gun, knife, etc.
3: He does draw when he decides at some point in the fight he is losing. It may not actually or legally be a true life or death fight, but he thinks so.
4: He does draw when he loses his temper inside the fight.
5: Dominant fervor. He draws after winning. He's essentially won but hates for the victory feeling and moment to pass. He further punishes the opponent by presenting a weapon and scaring him with his glee and threats.

Recognizing these five situational events should shape good training drills and scenarios.

What Should You Do?

Before, during, and maybe even after, when a weapon is drawn by you inside a fight, it can definitely stop or escalate the intensity and/or bring out even more weapons. The questions I am frequently asked are, "I live in a state where 'everybody' carries a gun, Hock. If I pull my knife to scare someone off? Or I pull my gun to scare somebody? What if he is carrying a knife or gun? Will this cause him to pull his knife or gun out too?"

Ahhh … well, yes. Yes, that can happen. In the same way that your words, your facial expression, your clothes, or even your stance can escalate an encounter. But, yes, that can happen. Should you always pull your weapon with the first blush of a problem? Automatically? No. The problem must percolate to the level that reasonable and prudent people think it is justified. Police deal with this pressure almost on a monthly basis, or maybe a weekly basis, and in some tough places maybe even daily? It's an acquired skill. A feel.

"Should I always throw the long pass or always hand off the ball to the running back." No. I can't answer that on paper or at the lectern. Not even Tom Brady can. How could we? It is situational. It is best to have a few handy plays up your sleeve and wing it. (Well maybe not as many as Brady has up his sleeve, but a fella' needs some tricks.)

So, I simply cannot answer that hypothetical question with a "do-don't do." It's a "call." A call you must make in the moment just like a quarterback. HIKE! What's the field look like?

I would like to start a list of very specific situations to help out in the decision making, but then this little essay would grow to textbook size.
- There /Not There. (Why did you go there? Why are you still there?)
- Draw/Don't Draw.
- Point/Don't Point.
- Shoot/Don't Shoot.
- Leave/Don't Leave.

Still, part of your decision-making is based on what you see and think and how well you are trained to think and see. This brings us right back to the "who, what, where, when, how, and why" questions I have used as a foundation for decades now on just about everything we do.

So often people want a quick, magic bullet answer. There is none, and I'm sorry; I have no magic bullets like this for you. If anyone is selling you a box of magic bullets? I wouldn't buy them.

Addendum 7: A Final Word. The Knife Training Business, Then and Now

For a while there was a lull in knife teaching, perhaps because people became fascinated with guns as they became more widely affordable and available. And for social and political reasons for many years knife training was embedded in more socially acceptable "Martial Arts'. Then in the 1990s there was a "resurgence" if you will, a re-look, re-examination of knife material which stripped away the art and focused more on the actual tactics. Some called it "knife fighting," but I don't like that term. But you are still indeed, fighting with a knife when you are…fighting with a knife. Still, I don't like many terms, images, messages, logos relating to the knife and knife fighting. By that time in the 90s, I was in police work for quite a while, both in the Army and in Texas, most of that time as a detective. I'd seen and experienced working on a lot of knife crime, as in aggravated assaults, rapes, attempted murders and murders. I have been attacked by a switch blade, a shaving razor and a small and a big ax.

I know the depressing, dark side, the wet side in juxtaposition to all the smiling people having fun, slap-dashing around in gyms playing tag with wooden and rubber knives. Knife training is often treated quite cavalierly. This doesn't have to be the case as the culture of pistol training is quite serious and full of foreboding and legal scares. Careful, mature training cultures do exist, and this must also become true in knife training.

In the early 90s, this edged weapon resurgence was sort of an international turning point in knife training. A reboot if you will? It first resurrected the old military knife courses and the semi-legendary names of yesteryear. They weren't "kuraty" superstars. A sophisticated look at them however, revealed, they weren't so sophisticated. So several of us, using the newer sports training methods of the time, and bolstered by years in Filipino martial arts or other historical backgrounds, stepped into the ring and made "new" knife courses. Gone was the martial arts uniforms, belts, etc. We wore jeans with pockets and regular clothing belts. Street clothes.

Some of the 90s knife pioneers? James Keating. Tom Sotis. Kelly Worden, Bram Frank, Bob Kasper, yours truly, to name just a few, but there really were only a few of us. (Vunak is an 80s pioneer.) We wore shirts, jeans and shoes. I even taught at times in a suit and tie. We didn't trust the old stuff, and we didn't trust the established martial arts either, even the Filipino applications of the knife are often tricky. (Do you want to walk around wearing a vest with 12 knives?) Be free. Think free. Be skeptical. Are you a replicator? Or an innovator?

Still, the old just helps the new. This was also part of a bigger "breakaway" from establishments that was going on in that decade. The world was seeing MMA (or at least ground wrestling) on TV like never before. And somehow a collection of old stuff, dressed in athletic pants, painted in "Israeli mystique" – Krav Maga – was really shoved down the throats of Tae Kwon Do schools as mandatory, by clever (and insidious) shaming, business groups, like NAPA in the 90s. Revolution was in the 1990s air! Jeet Kune Do was spreading into a heyday. Inosanto JKD/MMA was already doing Thai and ground, and so much more. Ever hear of "Shoot?" But, I guess the Israeli mystique was greater than the Bruce Lee mystique? Mystique? Yes. Ever so important in advertising, sales and manipulation. That's how we pick shoes, cars, purses and pistols (politicians, religions and… more) through manipulation. More on that later…

My knife course had a few odd, infancy names in 1990 and 1991, but it was quickly called "When Necessary? Force Necessary: Knife!" But the title was a little long and clunky and it was shortened to "Force Necessary: Knife!" I do prefer the longer, clunky name, as it completely explains exactly what I mean to say. Only use that force necessary when absolutely necessary. But I got around doing that knife material. Lots of traveling, lots of seminars. Even around the world. It lead to being voted *Black Belt Magazine*, Weapon Instructor of the Year and also being included in their BB Hall of Fame. I also "scored-very well" in the non-arts world.

 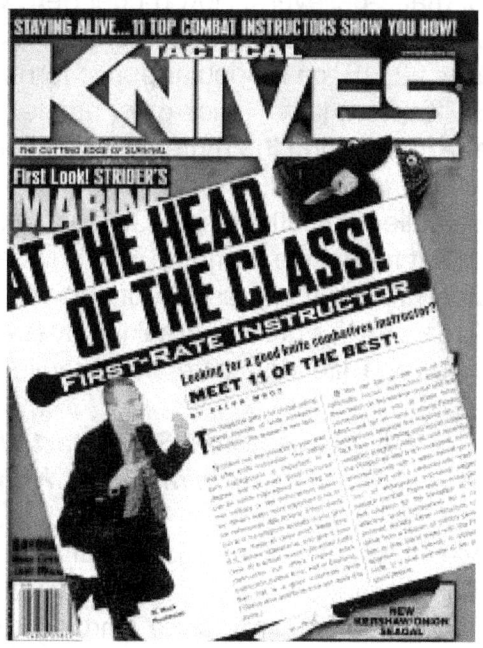

Black Belt. Tact Knifes. Halls of Fame. TRS. Such was the jargon and the martial/political stage of the 90s. Today, it's hard to grasp that the total, martial world communication back then that existed was with just 6 or 8 international, martial arts magazines. That's it! Try and list them. Yes, *Black Belt, Blitz, Martial Arts Illustrated, Inside Kung Fu, Inside Karate,* try and list them. They were the filter for us all. Talk forums developed slowly later and now, like the magazines, are almost all extinct.

Now? Nowadays, I don't know where the martial arts communication filter exists, specifically. The…web…the gazillions of webpages? The gazillions of podcasts? The gazillion of….Instagrams? Facebook? Yesterday's business card is today's webpage. And dipshits can pay to have amazing looking webpages. The battle for exposure takes a business up and down many extremely, frustrating, costly roads.

Remember this from the 1990s? When TRS and their classic advertising 'letter' campaign swamped us all in the land mail world and in EVERY magazine.

I did two films with them in the 90s and while we make fun of them, TRS did propell me into more international notoriety.

Me, Mike Gillette and Tom-The Arnold-Barnhart did some knife fighting sequences on the "steels," and they were not used, cut out, deemed to be too violent.

(We make fun, but I still get a check every January for over 20 years!)

Of course with all businesses, this 1990s knife movement kicked off a new interest and a fair number of new knife courses popped up through and to, by 2005-ish, often by less experienced, less organized people, and in my opinion doing less comprehensive programs. But this business evolution is to be expected. Invent a new "widget?" There's a knock-off. Then knock-offs with an "S." In the big picture of training and education however, not widgets, this is a positive thing. Awareness. Curiosity. Growth.

So when did true knife instruction get lost in a sea of make-believe wanna-be's? It happens slowly and then one day you are looking side to side and all around lost in a sea of make-believers. How'd I get here where the fake is nearly indistinguishable from what is real? Not enough Instagram pictures? Enough memes? Some 25 odd years later, in about 2016, on a popular public forum someone asked me what I thought of Johnny Swift's new, knife, quick-draw article. Of course it was named something super-spiffy like "Armageddon Instrument Production," but it's just knife quick draws. New, Biblical advice they preached, and published in the new amazing world of web-jargon magazines called something like "Organic Micro Evolution of Edged Prophetic Dynasty." (I just made that magazine name up, but how far am I off? Have you seen these seminar names lately? Are you impressed, or can you see right through the pretentiousness? So anyway, I read Swift's ground-breaking, testament as featured in "Retrograde, Skill Supremacy, Fusion Elite Magazine" and I replied on the public forum –

"Oh, I have to like Swift's article. It is virtually, word-for-word, from my 1992, Knife Level 1 outline."

My review/remark caused a lot of guffaws and a few smart ass remarks, among the 20 and 30 year old readers, most of whom were so submerged in modern "dynasty jargon" and up to their beards in mystique, and lost in the gazillion web world, they'd never even heard of us older guys from the 90s who pioneered these movements based on even greater pioneers before us who we credited with their work. I mean, who am I to comment like this on their latest fad-boy genius? I added that I was not suggesting that Johnny Swift plagiarized my outline, as it might have innocently been co-opted, or the older info has become so embedded into the "knife world" it was deemed as open knowledge. I get that. Sure.

I reminded the guffawers that the spread of education is a good thing and that at the very least, I only helped that. I said that the old just helps the new, and you have to remember the old, so bad history doesn't repeat itself. As Dave Spaulding likes to remind us, "It's not new. It's just new to you."

One guy was clever enough to say, "Well, sorry I missed you when I was 5 years old." I told him that was a pretty damn, funny retort. It was. But missed me? Dude, I never left.

But actually he never knew I was around to begin with. That is part of the mysterious "forgetting." (That level 1 outline is/was free to the public and has been distributed for literally decades, and my knife books have been for sale since about then too.) By the way, fad-boy Swift is already gone from the scene.

I added in that discussion with Mr. Wise-ass that the spread of education was a good thing, and I only partook in the process. Seriously, I frequently read as new, many old catchy terms/ideas/expressions I coined originally and published and advertised decades ago. They have literally become part of the modern language in this field. And it's not lost on me that imitation is the highest form of flattery.

My really big mistake in the knife world, training business is...I think, not emphasizing the knife training business only. Alone. My obsession was/is with covering the bigger picture. Hand, stick, knife, gun. That's "where it's at" for me (is that phrase too 90s? Yikes, maybe too beatnik 60s?). The 1990s evolved into the 2000s and my step-by-step into what I really wanted to do all along since the 90s. My goal is to create the best hand, stick, knife and gun courses.

```
HAND              STICK             KNIFE
STICK             KNIFE             GUN
KNIFE  >VERSUS<   GUN   >VERSUS<    HAND   >VERSUS...ETC...<
GUN               HAND              STICK
```

It's a mixed weapon world. Each subject I have is a carefully constructed 4-pillar, foundation. But I think when you shoot for this holistic picture, each separate pillar seems to get a little lost, a little less appreciated, a little less noticed. It also makes me appear to be less specialized. This isn't true. There's a big mixed weapon matrix: But anyway, back to the knife! Inside a comprehensive knife course is:

* Knife vs hand (NOT empty hand vs knife! That's unarmed combatives and belongs "over there."
* Knife vs stick.
* Knife vs knife.
* Knife vs some gun threats.
* Standing, kneeling, sitting and on the ground.
* Saber and reverse grip experimentation. (BOTH grips! You knuckleheads!).
* Skill developing exercises.
* Knife combat scenarios and situations.
* Legal issues and smarts.
* Criminal history knife stories.
* War history knife stories.

I do get a kick out of the occasional lame-brain who pipes up and says, "Knife training? Just stick the pointy end in the other guy." Especially when they spend about 10 thousand $$$ a year plus, shooting at gun ranges. Why not just stick the pointy end of the bullet in the other guy, too, Brainiac?

But, not focusing just on the knife is a marketing problem. I don't advertise or highlight "just the knife" like other courses do. In the marketing world of top webpages, indexing and emogi's this is a handicap, but in the real world of life and fighting, it is necessary to know and learn and apply all techniques comprehensively to be truly prepared.

No Flags. Oh, and I have no crutch system, no flag to fly, like Pekiti, JKD, Brazil-Mania, Krav. Silat. Arnis. Just little ol' me flapping in the wind. I can't draw in extraneous-system-people, as some of those are obligated to attend, even arm-twisted by "the system" they're in. Brand names are…brand names. You see flags change over time, loyalties change, times change, techniques and skills better survive the test of time.

No Mystique? Which leads me back to the first paragraph. We know the established advertising fact the "the grass is always greener on the other side of the street"? Other country really? The sewers of Spain. The temples of Thailand. The monasteries of China? The borders of Israel…the…and so on. Me? I appear to be just a bland guy with some info. I don't even have any tattoos! Long ago, I realized that tatoo you love today you'll want to cut off with your knife tomorrow as you yourself grow and evolve.

Plus, I avoid and dodge macho, death messages, grim reapers, and death images mystique. And I am not in a "mafia." I am a life-long cop. I fight the Mafia. I am not in any "cartel," or a "cult" etc. Look, I can make the distinction between something that is a little fun and ironic and something/someone that is sick and weird. It takes a little investigation too, to not jump to conclusions, but sick and weird is sick and weird.

Various other ultra-violent, whack-job messaging should be reserved as a primer mentality for very serious, military, combat groups. THEIR psychology. Their prep. Not cops and certainly not citizens. Mimicking them makes you look like a wannabe punk. Look at the lawsuits filed on cops and citizens. Go ahead, have a little death-engraved-logo on your cop gun and see what happens when you shoot someone. Have a patch or tattoo of a grim reaper with a knife, or a skull with a knife through it, and see what happens when you have to use a knife. We the police, the prosecutors search your history when you are in an assault, knifing or shooting. Mature survival, enduring the end game – as in the legal aftermath, is a big part of a well-thought-out, course. (Mature gun easily people understand this.)

And the serious military angle? Even with them, take a look at the most sophisticated, revered, respected, top-flight, Special Forces vets and most play it quiet cool like a gray man. Not like this silly fucker in New York for example – I read one New York City, very popular, international knife "cartel" group headline paragraph:

"I love it when I carve someone's balls off and put them in his empty eye sockets."

Shit man, you probably work in a fucking supermarket. And you think and talk like this? You need to be on a watch list. These idiots give us all a bad name. But images and expressions like this, or even near like this, this mystique, does attract a certain sick customer, usually young, or young in the brains anyway. (After my public complaints and comments on this, this moron took that line down.)

Lackadaisical about making rank and instructors. I don't really run the classic franchise business as seen in self defense, BJJ and Krav, other combatives courses, and Lord knows, classic martial arts. I am often lackadaisical about promoting people and making instructors. Other systems do this like precision clockwork, where I fail to emphasize this. It does hurt the proverbial martial, business model. In the same vein, I shun all titles like guro, grandmaster, sensei, etc. "It's just Hock," I say, which does not fly well with some organizations who base themselves on this structure. It's almost like I am insulting them? I'm not trying to. You do whatever you need to do to survive.

After the fall. However boring, I still do see some "knife people" all around the world. There are "normal" people, martial artists, historians, survivalists and hobbyists, gun people out there, interested in generic, evolved, knife material. There are. And that is who I mostly see when the knife topic comes around. Since I disdain the crazies and the fringers, they usually avoid me too. I know they know, I don't like them.

I always do a few hours of knife in every seminar and I do have the occasional knife weekend seminars when and where I realize I need to catch up with people's requests. And, normal people can always, sort of, hide their knife interests inside a classic martial arts name. To me the knife is inside of, part and parcel of, hand, stick, knife, gun crime and war, survival education.

So, me. Boring. No mystique. Not isolating the knife enough. Not promoting people fast enough. No skulls. No flags. No carved out-eyeballs. No macho. Here is where I differed from the others, perhaps even stabbed myself in the foot, in the knife training business, even though just a few of us are those innovator pioneers and turned the tide in the 1990s into what it all has become today. For better or for worse. Maybe you young fellers will learn from my mistakes?

It's always good to mention and/or thank your prior teachers once in a while. I always do. But, before you young knife guys make any sarcastic jokes about me again (and Kelly and Bram, et al?) Keep in mind…your modern instructors might have "peeked" at all my long, established materials, and would not confess. I might just be your grandfather.

Addendum 8: Knife Course Information

"This is my *Force Necessary:Knife* course. It is "painfully" simple. Pun is intended. I only teach you the very essence of knife combatives and knife fighting, self-defense. And as I mentioned earlier, I do hate the term "knife fighting." I have been training in this for more than three decades, and teaching "the knife" for more than two decades, well before it was cool - as some people think I was part of a "new' pioneer movement that brought the combative knife back in the 1990s. My revered system is taught all over the world, as far as China and Australia.

I have seen many knife fads and systems come and go. Many of them were and are shallow or crazy complicated and flashy. Many send the wrong message of sheer, mindless death and violence, ignorant or flippant about the harsh realities of the subjects of knives, crime and war. As a military and police vet, a detective-investigator of many knife attacks, robberies, aggravated assaults, attempted murders and murders I have a special view of the 'knife.'

Sorry, if you are looking for fancy, exotic, flashy, foreign, mysterious knife material with wacky names and weird people? We are *Force Necessary,* not Force *Un*necessary. In this course, we teach the following" - Hock

- Knife Vs Unarmed
- Knife vs knife
- Standing, kneeling, seated, ground
- Psychological issues
- Knife Vs Stick
- Knife vs gun threats
- Lethal and less-than lethal
- Skill, flow and speed exercises

Level 1: Introduction to Foundation, Ways, Means and Kinds of Knives. The Stress Quick Draw Module
Level 2: Support the Knife! While Holding. Introduction to the Dodge/Evasion Drill and Closed Folder Strategies and the Pommel Strikes Module
Level 3: Saber Stab Module. (Military "Quicker Kills" Module - Stabbing Part 1)
Level 4: Reverse Grip/Ice Pick Stab Module. (Military "Quicker Kills" Module - Stabbing)
Level 5: Saber Slash Module. (Military X Knife Fighting Part 1)
Level 6: Reverse Grip/Ice Pick Slash Module. (Military X Knife Fighting Part 2.) and Introduction to the classic Claw of the Cat
Level 7: Crossing Blades! The Dueling Module
Level 8: The Spartan Impact Combat Module and The Chain of the Knife Module
Level 9: The Death Grip of the Knife Module
Level 10: The Knife "Black Belt" Test

Further Specific, Intensive Studies

Level 11: Intensive Knife Ground Fighting Module
Level 12: Intensive Fist Knives/Push Daggers Module
Level 13: Intensive Knife Speed, Skill and Flow Drill Mastery
Level 14: Intensive Double Weapons, 2 Knives and Stick and Knife Materials
Level 15: ...and up: Upon Personal Development and Themes Upon Request

Teaching credentials:

1: Class Organizer
2: Basic Journeyman Instructorship after Level 3
3: Advanced Tradesman Instructorship after Level 6
4: Specialized Instructorship after Level 9
5: "Black Belts" on and after Level 10
6: Or just simply train for knowledge

MUCHAS GRACIAS, AMIGOS...

Starring the visages of:
Tom "the Arnold" Barnhart
Mike Gillette
Jason Gutierrez
Joe Hubbard
Rob Kloss
Jeff "Rawhide" Laun
Tom Pierce
Kelly Redfive
Roy Reynolds
Randy Roberson
Ronny Young
Scott Pedersen
Snake Blocker
Steve Lowery
Lyndon Johnson
Jim McCann

Photos by:
Mark Caswell
Jane Eden
Life Photo, North Richland Hills, TX
Ft. Oglethorpe Photography, GA

"Come on, get up, up, up, Tom, there's 41 more throws to go! Let's go!"

"You dirty bastard."

Seminars, Books and Videos: www.ForceNecessary.com

Get the **Training Mission** books and videos. They contain vital, foundational information for self defense, survival to support hand, stick, knife and gun. Plus, they follow the rank-by-rank, step-by step requirements of each course, including the knife training course progression. There will be a minimum, of six TM books through time.

Get and play these download, streaming knife training films.

Find them here:
https://shop.forcenecessary.com/us/C-Knife-Combatives/c/878

www.ingramcontent.com/pod-product-compliance
Lightning Source LLC
Chambersburg PA
CBHW051802100526
44592CB00016B/2525